# Hope Deferred
## Ingratitude, Disdain

MAINE-PATRIOT.com
3 Linnell Circle
Brunswick, Maine 04011


maine-patriot.com

Hope Deferred

"Brute force doesn't work with the laws of nature."
— *Ralph Ring*

# Hope Deferred
## *Ingratitude, Disdain*

# Contents

List of Resources
Other Publications

Hope Deferred

# Introduction
## *Ralph Ring Interview*

**Aquamarine Dreams:** *Ralph Ring and Otis T Carr.* A video interview with Ralph Ring, Las Vegas, August 2006.

*Also present: Gary Voss from The Ranch: a consortium researching exotic energy and antigravity systems.*

**Ralph Ring:** He says, *"You're gonna get onboard and you're gonna go some place and you're gonna come back. And that's all."* And he said, *"But I'm gonna tell you ahead of time, your brain will no longer...*

**Gary Voss:** *...be the same?*

**Ralph:** [laughs] *"You will lose it. Because it won't understand and it won't comprehend what's happening. So use your mind, use your feeling, come from your heart. Meditate. Go into a focus point and go to your higher thoughts and feelings, you know, rather than worrying about what is gonna happen."*

**Ralph:** *...these shutters are opening and shutting, creating all this reality you see around you. It doesn't really exist. It's all spirit. It's all energy. But we're creating this.*

Hope Deferred

# Start of Interview

**Gary Voss:** Do you want to give us an intro on how you and Otis Carr came together and what your background was at the time, and then bring us up to date?

**Kerry:** And how you worked with Jacques Cousteau?

**Ralph:** OK. Yeah. That's a good place to start. I got out of the service in '54. When I was in the service I was stationed on Guam. And they had the Korean outbreak and they shipped us out in the middle of the night to Korea. And I made the Inchon landing and I went through that situation, which was kind of unpleasant.

I got wounded about four times, and had frostbite and so forth. And I became very, very discouraged with the military. Totally. Because of — you know — everything. And I was an objector from the start on killing people, so I'd shoot in the air or whatever. They didn't care much for me.

While I was on Guam, the Marines would fight with the Army, and the Army would fight with the Navy — at bars. They'd go out and have fistfights. And I preferred to go down to the beach and look around and I finally got into scuba diving... skin diving, actually, with a snorkel, and found a whole new environment under water and I became fascinated with it. And so I kept developing that until the outbreak in Korea.

When I got back to the States I was in heavy weapons. I was with the 3rd Division, 7th Regimental Combat Unit, which is nothing more than machine guns, heavy mortars, and big artillery. Even though I put in for engineering school continuously in my time in the service, they kept putting me back in the infantry.

So when I got out I didn't have much to go on except my interest in diving. So I started a diving business, a skin diving, scuba diving business in San Francisco, and eventually graduated down to Southern California where my family was, my relatives and stuff, in Newport Beach and Costa Mesa, California. And I met my first wife and we got married and had a couple of children.

And I was diving on my own then. I was doing a lot of diving. I was doing abalone diving, research and development diving, and recovery diving. And it was going along well but my wife didn't like the idea that I'd be away for a couple of days on the boat, and stuff, 'cause the kids were growing up and they needed their dad, and so forth.

So, to make a long story short, I went to work for the manufacturing plant, at that time, for US Divers was located in Costa Mesa, or Santa Ana. And so I went to work for US Divers which had developed the SCUBA, you know, Jacques Cousteau's scuba gear. And right away we hit it off because I am constantly a researcher and developer myself.

So I went into the research and development department and we'd take trips out to Catalina. My job was to test the wrap-around masks, at one stage.

Anyway, I got really involved with that, but we'd stay out longer and longer on research trips. And my wife was getting very insistent that I get something a little less dangerous and a little more domestic at home. [laughs] So she found this ad in the paper and said, *"Advance Kinetics in advertising. They need lab techs, laboratory technicians, and research technicians. Why don't you go check? Because your interest has always been in science and you're always 'off.'"* (When I get home I'm always tearing things apart and stuff.)

So I went over there and it was lunch time. Everybody was out to lunch. So I was walking down the hall and I passed the Director's office, Dr. Weinhart. And he says, *"What are you doing?"* And I says, *"Well, I'm looking for a job; you had an ad in the paper."* And he says, *"Well, they're all out to lunch."* And he says, *"Come in. Let's talk. What have you done and what is your background?"*

*"I don't have credentials except bumblebees and lizards and things that I've studied and I've found out that there's quite a bit of credibility to natural law that I apply to things and it always works."*

So he said, *"Magnetics? You're interested?"*

I said, *"Yeah. I've studied magnetics all my life. I love it."*

*"Well, you know, coincidentally, the guy that was working our magnetic project just left."* And, *"Come to work tomorrow morning. You're going on the magnetic project."*

And I said, *"Fine. Great."*

So what it was... I had a bench, a workbench and there

was a cathode ray tube shooting, firing electrons (and I had an oscilloscope mounted with camera, high speed camera, high speed everything) through a magnetic field. I was firing electrons. And he said, *"Take pictures of them. The idea, your goal, is to get one electron completely through the field without deflection, without it being pulled to the positive or negative."* I said, *"Fine,"* you know. *"No problem. It's an easy job."* So I just kept taking pictures - quite extensive, and expensive. Every day, you know, it was about $1,000 worth of work every day that was paid for by the taxpayers for the research. And I started questioning it. And my affiliation with nature told me that they were using force.

**Voss:** Brute force.

**Ralph:** Brute force. And it doesn't work with the laws of nature.

**Voss:** No. It doesn't.

**Ralph:** So I said, *"This is never gonna work. I can appreciate this guy leaving. He got fed up."* And I was getting there too, fast. So I went home. And I had gone to garage sales and collected, you know... I had an audio amplifier. I had a frequency generator. I had different things at home, and I tore apart a TV and got a cathode ray. I started the experiment on a small scale on my living room floor. And I set it all up and got everything the way I felt it should work. And, instead of forcing the electron, I pulsed them. I just gave 'em a pulse. And that's all I did. And they, on their own, started a circular motion.

**Voss:** Traveling in their own pattern, and how much they wanted to at a specific moment.

**Ralph:** Yeah. And they went from negative to positive, all the way through to the end of the...

**Voss:** Just feeding it back to the source.

**Ralph:** And I said, *"My god, that was simple."* Because the first shot went through. And then I did many, many more and they all went through without deflection at all. So, I'm happy. This is gonna get me a raise, maybe.

So, then the next experiment: On the bench next to me they were working on levitation.

**Voss:** Who's "they"?

**Ralph:** Other technicians, other engineers were working.

**Voss:** And what department were they?

**Ralph:** Advanced kinetics. The laboratory was huge and we had different work benches. They were working on lasers to the moon, levitation.

**Voss:** So there were different interests involved in some of the projects as well.

**Ralph:** The government was funding this. This was all government funded research.

**Voss:** Department of the Army?

**Ralph:** I don't know.

**Kerry:** To get back to your story, though. So you had developed this pulsing, and you're saying, next door...

**Ralph:** The next bench over they were working on levitation. And they had... just coils... you know, iron with copper wiring. And they had steel balls and they would put them on top and fire it up and they would levitate the ball for, I think, 4 to 8 minutes, and it would burn the coil out. They were called "igliotrons", I believe, was the correct term for them. And they'd have to go to supply and get another one, and get another one. And they were burning up two or three or four of these. And in those days (that was the '50s) they were like $400 a copy. And they were, just, *"We don't care. We've got plenty of 'em."* And they were burning these things up.

So the other experiment I did at home was, I took a 15-inch woofer speaker that I got at a garage sale, or, I don't know, out of a sound system somewhere. And I put it just flat on the carpet on my living room floor and attached my audio amplifier to it and I started experimenting with, like, acoustical levitation, thinking, you know, they were using this force to push up, and they were using a lot of power. I'll try *sound waves,* I'll try *sympathetic vibrations* or whatever.

So I fooled around with different objects and I'd have tentative results. They'd start pulsing and stuff. But then I put a ping-pong ball in the center and I kept fooling, and I think it was at 28,000 cycles I got the ping-pong ball to come up.

**Voss:** Interesting that you mention that because I recall seeing a news clip back in 1989 showing that scientists "discovered" how to do exactly what you just described.

[laughter]

**Kerry:** Which you had done, in the '60s or something? The '50s?

**Ralph:** The '50s. The solution was quite simple. It could be done today. It could be duplicated today I suppose. I've

Hope Deferred

never tried it. I didn't need to go back to it. But it was a very simple operation because *you let nature do all the work. And all I did was understand what was happening.* So, the thing with the ping-pong ball, once I got it to levitate I was excited as heck. And my wife said, *"Come to bed. Come to bed."* So I went to bed and the next morning when I woke up the ping-pong ball was still sitting there.

**Kerry:** Levitating? That's amazing.

**Ralph:** Levitating. I think it was 28,000 cycles.

**Voss:** No heat?

**Ralph:** No heat.

**Kerry:** Just sound?

**Ralph:** Acoustical sound. That was all it was.

**Voss:** Was it audible to the human ear?

**Ralph:** No. I couldn't hear a thing.

**Voss:** So you're talking perhaps high frequency levels, or ultra high?

**Ralph:** Yeah. Yeah. You know, I experimented down in the lower ranges and nothing seemed to happen. It would bounce and stuff. But when I got up...

**Voss:** In the UHF.

**Ralph:** [nods] Yeah. Then it happened. So I thank my wife for pushing into that direction because this is my field. This is what I always wanted to do. And I think, *"We're on to something here."* And I could put my two cents into the pot and help humanity do something.

**Kerry:** Way back then, you knew all this?

**Ralph:** Oh yeah.

**Kerry:** And so you took it to these guys, right? And how did they react?

**Ralph:** I took it to Dr. Weinhart himself. I took Polaroids and I wrote up papers on it, just like I did at the lab and I took it in to Dr. Weinhart. And he says, *"Close the door. Come in. Sit down."* He looked it all over and stuff and he said, *"Yeah, I know it's that simple, Ralph. I know that. I know that. But this is a government-funded research facility. We count on the funds to keep us going. We're not necessarily interested right now in finding the answers. We're interested in looking for them. And we get paid handsomely for looking for the answers."*

And I said, *"Well, here, look. This works. I mean, maybe I don't know what I'm doing and maybe it's not right, but I thought if I turned it over to the boys here we can come together. And this is a lot simpler than $400 a copy for igliotrons and wasting our time with the cathode that you've got set up."*

And he said, *"I can appreciate what you've come up with. And I didn't think you'd get there this fast with this because of your interest in natural law, but I'm gonna have to shred this."* (He had a shredder right there.) *"I'm gonna have to shred this and tell you to go back to work on what you were doing."*

Well, right there my whole world collapsed. I mean... I thought, *"Where am I?"* My whole attitude, my whole demeanor toward the world changed.

**Voss:** Yeah. Who are these people anyway and who are they really working for?

**Ralph:** Right. That's exactly the way I felt.

**Kerry:** So eventually you actually left the job, right?

**Ralph:** Yeah. To make a long story short, I went back, I worked another two weeks and I couldn't stand it.

*"That's it. I'm done."*

But during this period I started meeting people, just coincidentally.

Most people I talked to outside of the lab didn't want to talk science. They wanted to talk other things, so I had very limited contacts with other people who were interested in science. Except I met this one person who said, *"Well, you know, what you're talking about is exactly what they're talking about in these meetings that I go to. And the name of the meetings are called "Understanding". And they were developed by a person called Daniel Fry who was in the UFO stuff. And they want to understand more. Why don't you come to one of our meetings and talk?"*

Well, I went to the meetings and kind of duplicated what I just said about where I was working, you know, and they said, *"Oh, oh, you've got to meet somebody. You've got to meet Mr. Carr."*

**Voss:** What year was this now?

**Ralph:** I think it was late '59 or early '60.

And they said, *"You guys are talkin' the same thing. The same exact thing."*

And I said, *"Well, OK."*

And he said, *"Well, coincidentally he just had some bad luck in Norman, Oklahoma."*

(That's where he was trying to demonstrate the craft, you know, the flying disks. And they started negatively defining his work. And the newspapers got ahold of it... "He's trying to get funds to do something that's impossible." And: "Science has just never heard of such a thing." And so forth.)

*"So we're going to bring him out here and we're gonna get a lab out here, with his entourage. And let's go, let's try another place, another time and see if we can get somewhere."*

So they did.

And I met with Carr and his entourage. He had Dennis Ripolte, Norman Colton, Wayne Aho. I don't know. There were about six of them.

**Voss:** He had a little consortium going.

**Kerry:** And where was this based now? Where was your group meeting?

**Ralph:** This was in Costa Mesa, California, where these "Understanding" meetings were and that's where I met Carr. They found out they were after Carr. He was having all kinds of misses. They were trying to quiet his efforts.

**Kerry:** When you say he was having misses, actually people were trying to kill him?

**Ralph:** Yeah. They were threatening and then, you know, he'd have to be very careful where he went because he'd find people kind of, you know, very curiously observing him, and you know, things like this.

**Voss:** They already knew what you were up to and they probably were following you as well as following him.

**Ralph:** That's a good point and I didn't bring this up, but I think it's important. You've heard by now... it's all over the place... about three Men in Black?

**Kerry:** Right.

**Ralph:** OK. This was back before I'd even heard of such a thing. These three guys showed up at my door after this experiment and after Weinhart had destroyed these things.

Honest to god, they were in black suits. [laughter] And they said, *"We're from the DeWalt School of Electronics and we've heard about you. Can we come in? We want to know about your experiments and what you're doing and everything."*

And I was a little hesitant but I invited 'em in and I started talking.

And my wife said, *"No. No. These guys don't feel right."*

**Kerry:** Ah ha!

**Voss:** She's very intuitive. She had a bad feeling about them.

**Ralph:** Yeah. She sure did.

And I said, *"Well we can't kick 'em out."*

But they became a little more insistent.

Like: *"Well, give us how you did this."* And: *"I want the details,"* and stuff.

And they're not giving me anything back. They're just kind of taking.

And she caught this and she goes, *"I'm going to have to ask you guys to leave right now. You can come back later or whatever you want to do, but you've gotta leave right now."*

And she kicked them out of the house.

**Kerry:** [laughs] OK. So you started in Costa Mesa. And didn't you move out of there, or something?

**Ralph:** Yeah. The "Understanding" group had a cabin. There were lots of people in the "Understanding" group. There was, I don't know, a couple of dozen people that would meet... had a cabin up at Lake Arrowhead, which is down by the riverside, up in the mountains in California.

And they said, *"We've got to get Carr and you guys in a safe place. And there's a nice big cabin, and room enough for everybody. Go on up there and then we'll keep workin' on what we're gonna do."*

So I got up there, talkin' with Carr and his protégés, his people that he had with him, and man, I just lit up like a Christmas tree. I mean, I was on Cloud Nine! Man... he was answering questions that I had on things and I was answering questions he had on things and it was just... Man!

**Voss:** Connected on all kinds of levels.

**Ralph:** God! - It was the most wonderful time of my life. We were feeding the raccoons to keep our minds... We were so anxious to get goin' on the project. And we had a phone call and they said, *"We've got you a place. It's just down the hill from where you're at, on the other side, in Apple Valley, or Hesperia, California."* Near Victorville. It's coincidentally, because all these people moved on feelings and spirit, if you will.

**Voss:** This is the same era, I wanted to point out, that George Van Tassel was having a lot of UFO meetings out at the Integratron near Joshua Tree.

**Ralph:** I'm glad you mentioned Van Tassel. I had forgot. I had ordered from Europe Tesla's big book and it did get to me. And I was going through all the patents and everything in the big book. And when this thing happened with my wife kickin' these guys out and everything I got a little apprehensive. And I decided... I knew of Van Tassel. And I knew a little bit about his background. This was before I met Carr. So I took a trip. I got in my car because I was going to try to meet people that...

**Voss:** Would be more accepting?

**Ralph:** Yeah. Who were more accepting. And I took this "Bible" down to Giant Rock, Joshua Tree, California, and met with Van Tassel and we had a nice talk. And I said, *"You know, I'm supposed to give you this,"* you know. *"I'm out of this phase of it. I don't know where I'm goin' or what I'm gonna do. This is it."* And I gave it to 'im. And I remember it was getting late that afternoon or evening and I went out and laid on a hillside and I looked up in the sky. And I saw hundreds, if not thousands, of whatever they were. UFOs. Spaceships.

**Voss:** Different, various shapes? Lights?

**Kerry:** Really?

**Ralph:** Yeah. Green lights or whatever. I don't know. There was hundreds and hundreds. They were coming over and they'd stop and come down and go up and around. And, *"Oh my god. This is really... this is really..."* And I said, *"What is this for?"* And what I got was*: "Because you did what you did."* Wow.

**Kerry:** It was like a kind of thank you demonstration of a sort. That's amazing.

**Ralph:** And I just chilled all over.

[ *"And suddenly there was with the angel a multitude of the heavenly host praising God . . ." — Luke 2:13* ]

**Voss:** "If you build it, we will come." [laughter]

**Ralph:** Oh, man! So then I got back and they had set up the meeting with Carr and then we got down into the laboratory down in Hesperia, down in Apple Valley. And we started setting up shop.

We had a little machine shop set up and we had, you know, all kinds of stuff to do things with, but we had a couple of models that they brought with them that were semi-operational. So the first experiment that I saw that just knocked my socks off was... We set it up on one of the work benches and attached — not electricity, but sound waves, if you will. Or maybe it was, I think... I'm not sure.

Anyway, this was a small model, about, was it two feet in diameter? Two or three feet in diameter. And they said, *"Well, take a look at this."* So they fired it up. Hardly any noise, just a hum, a vibratory hum. And it was made out of aluminum. I touched the surface of it and it felt good, but I could feel the vibration. And so they kept increasing the energy and then there was this feeling... Jeez, it felt like somebody opened a door and a cool breeze was coming through. It felt really good. And at that time I went to touch it again and it was like jello, it was getting soft, getting really, really soft, like I could put my fingers through it. Better than jello, because it didn't stick or anything. I put my hand in and pulled it out.

**Voss:** Oh my goodness! And what did it feel like when your hand was inside of the gelatinous material? Did you feel anything?

**Ralph:** Well, it was the same tingling that we were all feeling in this room. We had accelerated our efforts. It was like what it was doing, we were doing too.

**Kerry:** Oh, I see. So you were speeding up, kind of like in sympathy to the vibrations.

**Ralph:** Exactly. Exactly.

**Kerry:** That resonance that you talk about.

**Ralph:** Uh hu. And after the experiment, Carr... The way he briefed us on things was just, we'd sit down and have a cup of coffee, you know? It was just... He'd come out with *this wonderful stuff about the laws of nature* and how *that is our whole essence and if we ignore it, we're in trouble.*

He's got to understand these laws and how they work for anything. *If you want a comfortable life, a good life, a happy life, and especially if you want to get anywhere in technology, you can't use brute force.*

And I told him about Advanced Kinetics and everything and he kind of laughed. He told me a lot. He worked with

Tesla. He had known him for a while and worked with him. And I guess by now you already know about the story of Tesla going to J. P. Morgan.

**Voss:** When he showed the wireless tower, *how to transmit power wirelessly,* he says, *"It's a real good idea, but how are we gonna stick a meter on it?"* [laughter]

The essence of *"We are in control."* It's really astounding. And he definitely sent the message.

**Ralph:** He said, *"If we go your way, Tesla, we'll have no more copper mills... no more lumber... and trees... for telephone poles... and wire."*

And Tesla said, *"Well, that's the idea! You can stick a pole 30 feet into the ground and 30 feet up into the air. I'll show ya. We can get electricity anywhere. It's all around us. We're living in it."*

And Morgan said, *"No way, Tesla! There's no money in that."*

**Voss:** J.P. Morgan, from what I understand, was also one of the first, one of the pioneers in the military-industrial complex. He was THE man. And soon afterward, he pretty much picked up the Bat Phone to Washington and said, *"Hey, we've got this 'loose cannon' on our hands,"* and the implications of the conversations pretty much took care of burying Tesla from thenceforth.

And, I guess, from what I understand from reading some of the journals, they gathered up all his equipment and shipped it off to Wright-Patterson Air Force Base, and I guess they put him up in a hotel, the Waldorf Astoria, and gave him a government stipend. And the agents were always on the crawl, prowling everywhere, interrupting his conversations and pretty much filtering out any connections with the outside world to him.

**Kerry:** Did Carr mention what happened to Tesla? Did he talk to you guys about that?

**Ralph:** Tesla became discouraged because of the lack of interest in, you know... I mean, he'd take 'em a new idea, bring out a new idea, and show 'em the simplicity, that there's nothin' to it. And they'd say, *"Well, there's no money in it. Forget it."* I mean, everything he'd bring up, you know...

**Kerry:** So it was Carr's point of view that Tesla was discouraged. But did Carr sort of relate his being hounded, you know, by the military, or being shadowed and so on? I mean, what happened to Carr as being the same thing that happened to Tesla? In other words, did he talk about that at all? Carr? Before he died?

**Ralph:** Well, I guess. I don't know. Carr didn't talk too much about the threats or anything that Tesla had, you know. But I was under the impression talking to Carr that there were many, many things happening, that were trying to keep Tesla under wraps, trying to keep him quiet. And Tesla had told him at one time, he says, *"You know, I may never make it in this generation to get these ideas out. This is all just free energy. Free."*

You've got four elements: sun, water, the air, and the earth. They're all free. They have been, forever, and they always will be. And we're not using 'em. We're inventing ways to put meters on 'em and sell 'em. Even selling air at one time, and now they're even selling water!

**Voss:** Who would have thought, hmm?

**Kerry:** [laughs] Yeah.

**Ralph:** So Tesla told 'im: *"All this that I'm sharing with you..."* (And he thought Carr was brilliant. He thought, you know, he was grasping everything that Tesla was telling him, because Carr had been into nature for years himself.) He said, *"If I don't make it,"* or *"When, I don't make it, because I probably won't make it, you take it and pass it on. And if YOU don't make it, pass it on."*

But it's going to get worse, because they've already challenged nature. Man, way back there, had challenged nature. *And what goes around comes around.* Natural law: It will come back on us.

**Kerry:** Basically Carr did exactly what Tesla asked him to do. He took it forward.

**Ralph:** Yes.

**Kerry:** In a sense you are taking forward what Carr...

**Ralph:** Oh, you bet...

**Kerry:** You seem to be the person that is like a descendent of Carr. In that line. Am I right?

**Ralph:** Yeah. I would say so.

**Kerry:** It's so amazing to me that you're so unknown.

**Ralph:** Well, there are many reasons for that.

**Kerry:** Because we would like to actually know why you're so unknown. You know what I mean?

**Ralph:** OK. I will tell you. They were all hit-and-miss. But Carr was always on and I'd stay up all night. We'd be looking at the stars and talk all night and never need any sleep. I mean, I'd go to work the next day and just feel pumped up all the time talkin' to this guy. He was just... wow. And I said, *"You know, don't worry. We're gonna get this thing going here."*

And they said, *"They're closin' in. They know he's in California now."* And some of our experiments on some of the craft... We operated different principles. And some of them would create a corona on the outside of the... We'd operate these little...

**Voss:** The dielectric principles. The ionization process.

**Ralph:** Right. And so even though it was daylight sometimes [makes a swish sound of object moving fast] you'd see these things. And the people in the valley were...

That was the era of flying saucers and stuff and so they thought, *"Oh my god, this place's got flying saucers around here,"* and stuff; well, that and the fact that the "powers that be" that were trying to reduce Carr's activities were following him, trying to find him. So he said, *"We're just gonna have to keep workin' on this."*

Carr had made arrangements and I went with him to meet with a representative of the General Motors car company, I think, at Riverside. The guy committed to meet with Carr, 'cause Carr told him a few things that interested him. So I went with him. And Ripolte was there. And I think Aho was there.

And in a very precise way, Carr said, *"You know, we can levitate these machines now. We can get off the Earth. We're killing a lot of animals. We're destroying the plant life."* He said, *"Within a year we'll have these things going. We can* start with the automobile. It's obsolete. We can get these things going. And then the homes."

(Which is my interest. I've always wanted, you know, why not? Like the Jetsons, for instance, you know, *floating homes*. And then maybe cities, and then maybe countries. Who knows the end of it?)

But this guy got real, real aggressive and said, *"You put 'em up there, Carr, and we'll shoot 'em down!"* That was his words. *"You put 'em up and we'll shoot 'em down!"*

**Kerry:** Wow!

**Ralph:** And I was flabbergasted! Like... why?

And he says, *"You're advocating an energy field where there's no money involved. We can't..."*

**Voss:** We have no means of controlling it, was his premise.

**Ralph:** Right. *"You're pulling energy out of the air which is all around us and using that to transport, or teleport, or whatever..."*

**Kerry:** So you guys just basically walked out of there and said, "OK." What did Carr say? Did he say, "OK. I won't do it any more?" What did Carr say? I'm curious, after that... Sort of a standoff?

**Ralph:** No. Oh, I don't remember his exact words, but he was very, very good, the way he said it to this guy.

**Kerry:** Oh really?

**Ralph:** Yeah. He said, *"It's only a matter of time until it comes back."*

**Voss:** "You can't stop us. You can't stop IT."

**Ralph:** Yeah. *"It's here. Whether it's here today or tomorrow, it's here. It's rapidly approaching the point where is HAS to be. Not 'want to be,' but HAS to be."* And he said, *"I'm sorry you don't see our point because we were willing to work with you. You could find us and we could show you what we can do."* And the guy wasn't interested. So we just left. And that was, you know, another acquaintance that I had with *the system that we live in* that, you know, I could never accept.

We went back to Apple Valley and said, *"Let's get the 45-foot craft going. Let's have people onboard. And we'll document it and have the proof and then we'll have the "Understanding" group finding ways to let people know what we were going to do. We were going to have live demonstrations eventually."*

To make a long story short, we went through stages and we finally got to the big craft. [film shows technical drawings of various craft] There was actually two of them but there was the one craft that we were ready to try an experiment with.

**Kerry:** And how big was that, again?

**Ralph:** 45 feet in diameter. And we were by that time...
You know, we didn't have any fences or anything around us and you could see the thing from the road, and stuff, and we knew it was only a matter of time before the looky-loos started getting there. But we didn't care. We knew we had to do something, because now General Motors was gonna go tell whoever what we were proposing, it wouldn't be long before they found out what we were doing.

Well, he said, *"Come on. We're gonna go."* He got us in the briefing room and he told three of us... I don't remember who the other two guys were now. It wasn't Ripolte. It wasn't Colton. It wasn't Aho. But there were three of us.

And he said, *"What you're gonna do, you're gonna get onboard. We're gonna go downrange."* (We had a 65-mile range in Apple Valley. Where we eventually wound up was about 10 miles, I think, down range.)

He said, *"You're gonna get on board and we're gonna go some place, and then we're gonna come back, and that's all."* And he said, *"But I want to tell you ahead of time your brain will no longer..."*

**Voss:** Be the same?

**Ralph:** [laughs] *"Well, you will lose it, because it won't understand, and it won't comprehend what's happening. So use your mind, use your feelings, come from your heart. Meditate. Go into a focus point and go to your higher thoughts and feelings, you know, rather than worrying about what was gonna happen."* So he said *"It's gonna be a strange experience for you, but it will happen; we'll document it."*

And so we got onboard. And what it was there onboard was just like *a small crystal ball* in the center. (It wasn't actually in the center. It was off center a little bit.)

And it had a... I think it was a laser; I don't know. But there was a white light coming up from the bottom of it, shining up through it. And it just beautifully broke the color spectrum from infrared, red-red, orange-orange, yellow-yellow, all the way around 360 degrees. Anywhere you wanted to go, any degree you wanted to go in, the color spectrum was there. I thought, *"Boy. That's beautiful!"* And we'd been briefed on it, but until I saw it I didn't realize what was going on.

And he said, *"OK. Just relax. We're gonna go to an area that symbolizes..."* (He used to use symbols a lot. I mean, he'd say, you know, *"Talking is useless. You have to use higher-than-talking symbols to reach the mind."*)

**Voss:** ...thinking in pictures.

**Ralph:** ...in pictures. Right. In fact, off the subject for a minute, when I used to read a lot one of my greatest people was Kahlil Gibran. He wrote the book, *The Prophet,* and in there one of his sayings was, *"Half of what I say to you is meaningless, but it's necessary so that the other half may reach you."*

And I thought, *"Oh, now I get it. You have to come from the soul or the heart, or it's no good. It's just goin' 'round in circles."*

**Kerry:** Was there something about choosing the blue spectrum in order to...

**Ralph:** Aquamarine. We were in touch. I don't know if we had walkie-talkies, but I remember that we were in touch, and he said, *"OK. We're going to aquamarine. That's over there. [gestures right] Hang on, boys, let's go."* So we set there. I'm just doing this from memory...

**Voss:** So you're all *collectively focusing the same collective thought,* to feed this energy into a center focal point which is the ball.

**Ralph:** Right. And this crystal ball, then, started just closing down [with hands demonstrates a sphere getting smaller] and focusing on aquamarine. The whole ball became aquamarine. *"My god, how did he do that?"* He told us later that we were a part of doing it because we were focused on it. I thought, *"Oh, oh, oh, this is great!"*

**Voss:** Like a biofeedback mechanism is synergistic.

**Ralph:** Yeah! So we got focused on it and then I was waiting for the thing to move now. And nothing seemed to happen. And then Carr said,

*"OK, boys, get out of the craft and see what's goin' on."*

*"Didn't it work, or what happened?"*

He said, *"Come on. Get out of the craft."*

We got out and we were *down range* 10 miles where this aquamarine area was.

**Voss:** I'm guessing that this whole process, you're talking about probably a few minutes.

**Ralph:** Oh, yeah. Yeah. I'll get to the timing in a minute.

So he said, *"All right. Pick up rocks. Put 'em in your pocket. Take some grass or whatever you can find. Some tumbleweed. Whatever you can find and get acquainted with where you're at. Because when you get back you're not gonna remember any of this."*

That was the gist of the whole thing. So we did, and we got back on board and then [makes swishing sound of fast movement] we were back.

And we got out of the craft, went in to debriefing, and said, *"Well, what happened? It didn't work, did it?"*

*"You don't think it worked? Check your pockets."* And so we checked our pockets. And here's these dang rocks. I had grass stains, I had everything. I said, *"Oh my god!"*

**Voss:** But you didn't have memory of this?

**Ralph:** No memory. No memory at all. I remembered *later,* being there and picking up the rocks. It was just like...

**Voss:** Like it was a dream or something.

**Ralph:** Like it was a dream. Exactly. You advance your imagination to a point and then you'd forget about it. And so I thought, *"This is the most incredible experience I've ever had."*

And he said, *"No, no, it's simple. Your brain is there to operate your body. You're in a vessel here. It's an illusionary vessel that people don't realize because we're creating it in microseconds. From one second to the other these shutters are opening and shutting, creating all this reality you see around you, but it doesn't really exist. It's all spirit. It's all energy, but we're creating it."*

And he was blowing us away.

But he said, *"Your brain has a capacity limit. It goes to a certain point of its responsibility and unless it's in touch with the Mind, unless it consents to be in touch with the Mind...*

**Kerry:** The Greater Mind.

**Ralph:** Yeah, that's the Mind of all of us, because we're all One. *"Unless it gets in touch with that, it doesn't know what's goin' on."*

[ *"It is the prerogative of the ever-present, divine Mind, and of thought which is in repport with this Mind, to know the past, the present, and the future." —Science and Health 84:11-13* ]

**Voss:** Am I to presume that at that moment that you had that flight of 10 miles distance, your brain was being stretched like a rubber band, but when you went back, you went back faster than the memories of the experience could come back and your brain could realize it?

**Ralph:** Yeah, something like that. Yeah.

**Kerry:** I don't know. Days later, months later, you could remember, like you said, picking up the rocks, then?

**Ralph:** Yeah, but I don't remember any movement whatsoever.

**Kerry:** You don't remember the craft moving? Or you don't...

**Ralph:** I'm sitting there and the ball turned to aquamarine, and he said, "Get out of the craft." We got out. There was motion, but I don't remember too much of that. I remember being outside. And then I guess we got back in and back to the base. But to us it was at least 15 minutes.

**Voss:** Normal time.

**Ralph:** Normal time. Yeah. I figured we'd been gone about 15 minutes.

**Voss:** So there is a time variation going on here.

**Ralph:** And Carr explained it. He said, *"Well, like, it's simple. People don't realize that Man in a sense created time. Time doesn't exist, in essence. It does when we create it and we have a beginning and an end to something. We call that time. But in a greater reality there is no time."*

**Kerry:** That's like the eternal Now.

**Ralph:** Yeah. We pegged it at 15 minutes and he says just a few seconds. We just went outside and back in time. I mean, *it's what you call it.* What you create is what it is.

And since then I've had experiences that have told me *just don't talk about it to anybody* because, you know, most people are not interested because they're tied up with the creature comforts and so forth. And a lot of people, when I start getting close to it, they get a little fearful because they don't understand. And they, of course, think I'm...

**Kerry:** What happened after? I mean, you made that test flight, right?

**Ralph:** Right.

**Kerry:** And so you didn't make that many test flights after that. Is that right? You guys got closed down somehow?

**Ralph:** I did the one test flight and then we did some things that were there at the plant. But we didn't go down range or anything because it was just about, within two weeks after that, that the FBI, and whoever these other guys, CIA, or whatever, came in to the plant. They came over with all their bells and whistles and said, you know, *"You're shuttin' down right now!"*

And we asked them why and they said *"Because of your threat to overthrow the monetary system of the United States of America."* That was their ploy.

**Voss:** Issues of national security, and what have you.

**Ralph:** Yes. *"And we're confiscating everything."* And they went into the offices, and they went into the lab, and they started just confiscating everything. And then they debriefed us and told us, in essence, *"You guys are wrong. You're attempting to overthrow the monetary system."*

**Voss:** And this is what we're going to do to you if you don't cooperate. Sign here...

**Kerry:** Well, did they have you sign something?

**Ralph:** No. I don't remember signing anything.

**Kerry:** And what about Carr?

**Ralph:** Yeah, they might have had him... he got really, really... His health started going fast after that happened, and I don't know.

**Kerry:** You were working on this night and day, pretty much, at that point? So you guys disbanded based on these people coming in?

**Ralph:** And they said, *"You are no longer allowed..."*

**Voss:** In no uncertain terms you will cease and desist !

**Ralph:** In no uncertain terms. *"We are watching you..."*

**Kerry:** So what did you do then? I mean, what did you do? Did you just go home? Did you try to work in secret at all at that point? Or anything like that?

**Ralph:** I tried to do it on my own, which I found out you can't do it. You've got to have other people.

**Kerry:** So, you and Carr, did you stay in touch after that?

**Ralph:** Well, they told us not to. Through understanding, I was in touch with Carr. We'll get together again. But he was really... He said, *"Nah. I don't think we're gonna make it this time..."*

# Q & A

The following interview contains questions inspired by the members of Ralph Ring's website:

http://bluestarenterprise.com

Hope Deferred

# 1
## Some have referred to the OTC-X1 as your craft. Is it your craft?

Is it my craft? No. That's as far away as you can get. No. The OTC-X1 belongs to everybody. The conception was put together by some very meaningful and caring people. And it was based on very caring and intelligent beings that wanted to see this happen.

# 2
## Do you have any operational teleportation craft?

At the present time? Well, me, personally? We, I have to say that collectively, because we are working in many different areas, and we have reached the stage of the prototype reality where we are not ready to produce, and have been given the signal that they're ready to understand the technology in the reality that exists. Yes. The answer is, Yes. we do have something in the wings.

# 3
## Some are claiming to have heard you state that you already have several operational craft. Built without moving parts, and are using solid state components.

Well, the first part of the question... I don't recall the statement being made by me that I have these in full operation, or in operation at all. I did say that we were working on them, and we do have, waiting in the wings, things that are a tribute to making a better life for everybody.

And they're in essence, free of moving parts, free of pollution, free of maintenance of any kind, and we now have to raise the consciousness — that means, wake people up far enough to let go of the old ideas of aerodynamics and combustion engines, and so forth, and just look into the reality of an easier and simpler way that requires no fuel from earth.

The fuel comes out of the either known in our vernacular as magnetism. And its everywhere in the multiverse. We can never run out of fuel. And no moving parts, because in the '50s and '60s when we were designing and building them, we did realize to understand, in vortex mechanics and quantum realities, we had to counter rotate — this is the way nature does everything. It creates vortexes everywhere, and so we have no moving parts. Since then we've met engineers in the electronic field that have proven to us that solid-state circuitry is now alive and well, and moving parts are history. We no longer have to do them with moving parts. So digital solid state circuitry is an advance version of what we did back in the '50s and '60s.

4
**Can you define your role in the design and construction of the OTC-X1?**

Well now. My status was a lab technician. I left the previous job I had as a laboratory technician. When I went to Carr I thought that's were I would fit in. I became a gopher. I went for hamburgers — food and drink for the guys — I worked on the lathe, because as a kid I was a machinist. So I knew how to turn things. I knew how to use a bench press — a milling machine — and so forth. So I did cut some of the parts. I helped hold wires while they were connecting different things.

I was kind of a "day-Friday- boy" down there. So I had many, many jobs. I had multi-tasking all the time. But my job wasn't to design. Wasn't to build, *per se.* I helped them build it, but I was not the engineer. I didn't have the... even though I understood circuit geometry and how exacting it has to be or it won't work.

And I honored that while I was working. And everything I did was with that in mind. But there was quite a few of us working on it, so I was just a cog in a wheel.

## 5
## Can you describe your working relationship with Oris T. Carr?

Well, I love the guy when I met him. He was just pure spirit. His whole heart went out to humanity. He felt — when he looked and the earth — like I imagined — he conveyed to me, like from the viewpoint of Tesla, that we were slowly digging ourselves into a continuous wheel, that would just go around and 'round — and not evolve.

It would *revolve* but not evolve, because of the technology that was being used, based on a profit margin. There were the parts, the pieces... The ideas were based on, *"What is my return on my investment? What can I do... What can we do to make more money?"* And this was not Tesla's idea at all. When he was with Westinghouse and they designed the dynamos at Niagra

Falls — he did that — he was offered $50,000 dollars at one time, and he turned it down, and said, No, that's for our work. So that's the kind of man Tesla and Carr were. They put humanity first, and if the money comes it'll come. If it doesn't, we'll find a better way ; another way.

## 6
## Would your share with us the kinds of things that Carr would say to you as you all worked on the Craft?

He did emphasize to all of us all of the time, 24/7, the need for understanding what consciousness is — to be consciously aware of every move. If we removed a part from point 'A' to point 'B', we had to know exactly where we moved it from and exactly where we put it. And the idea of that training was to realize that in sacred geometry you have to have your alignments exact. If you don't put them in the right place at the right time, you'll get no results. So by putting them — and being conscious of where I put them, or where I put that, at all times, it falls into place, and eventually what ever your case is, will have an effect that's successful.

## 7
## What was the function of the large crystal?

It was to create resident frequency of a designated rendezvous or landing port down range that we were going to teleport to. Teleporting is something that has to do with being *outside* of what we call time and space. It is just a state of being — a state of the mind, and when the crystal resonated at the frequency equivalent to what was down range, then we simply had the ability to move through time and space and rendezvous at that point, and back again.

[ *"Then they willingly received him into the ship : and immediately the ship was at the land whither they went."* — *John 6:21* ]

## 8
**Do you recall who determined the frequency for the destination, and how it was obtained?**

The engineers took a truck and went down, and they had their instruments, and I don't know what they're called, but they equated the location to the light and color of the location as vibrating at a certain vibrations and frequencies equated to the light and color, so they said this was it. They scanned the area and said this is where the craft will come and go from.

## 9
**Due to changing seasons and other reasons, mapping the Earth's frequencies for the purpose of navigation and transportation would be impractical or impossible. Do you agree?**

Well, the answer to that is that there's nothing impossible... when, whatever is conceived by man, will be achieved by man. We weren't able to fly at one time. We weren't able to go to the moon at one time, and all these are history now, like spaceships will become.

## 10
## Did the 45 foot disk fly?

No. Flying is a misnomer. You really can't... Flying is based on aerodynamics, and the craft itself didn't fly. It levitated, and once it reached a resonant frequency where the intended destination was, it simply, to our physical reality, our eyes, would become invisible. It would go down range and come back. And, I don't know if I should go and deeper into it. ...If you like, I will.

## 11
## Yes. Please tell us more.

We were doing with — and Haarp kept us well aware of consciousness, and what... who and what we are.

Basically, we are energy — we are force-fields — all of us — so light beings. We've lost the awareness of this because we chose to use our powers of creation to create a 3rd dimensional reality — and move through time and space in segments.

Actually, energy cannot be created nor destroyed. And energy is everywhere. And we are part of the one collective consciousness that is also everywhere. So in essence, we moved down range at the speed of thought.

We had only to think of where we were going and we were already there because energy is everywhere.

## 12
## Do you recall seeing any plans for the craft?

Well I saw a lot of different plans — and we had scores of plans that weren't... We were obsoleting a lot of the plans. Because some of the things we were working with didn't come up to speed — didn't come up to the qualifications we required, so there were plans — I might say — needily stuck away. But they were on the benches, and they were in the offices, so there were a lot of plans around.

13

## Can you describe a typical work day at OTC Enterprise?

Well a day at OTC Enterprise was a 24 hour day.

Someone was on benches, or on the presses, or on the designing tables, using what we had to put things together. So... and I remember being there a 3:00 o'clock in the morning when there was almost the whole crew working. But it wasn't a typical 9-5 operation. Everyone loved what they were doing. And that made the big difference.

When I chose to go down and take a rest, I couldn't rest very long because I wanted to get back to what I was doing. That's how strong the love of the work was. We wanted to see it work. And the other people, the other Engineers and the PR people and staff, were the same way. They'd get an hour or two of rest, and not even that, and they'd want to get back to work. It was a wonderful feeling. We loved what we were doing. And every time we made a step, we'd all congregate around and Carr would help us understand how that stuff came about and what it was, and how important it was to be in alignment with natural law. Keeping resonant frequencies with the flow of energy, is it flows.

## 14
## What you're describing sounds very similar to what's happening in the Pods today.

Yes indeed. And they're very... as anxious now... which does my heart a very very good feeling... because they're as anxious today as we were in those days. They see... they understand the geometry, the codes that Carr put in the *"Dimensions of Mystery",* a book of poetry and prose that he wrote.

When he was turned down by the patent office to patent a devise with levitation, he had to pull that out, and he wrote a book and said, when the day comes that they will be able to decode this, we'll have space ships again.

For this is the sole story and simplicity of levitation and how simple it is.


## 15
## Getting back to the topic of blueprints, have any survived that fateful day of the raid?

Well, yes. There were remnants, and I've carried them for years. Some of the sketches, some of the specs, and a few of the parts, but for the most part, realizing that the System was not ready for what we had to offer, we knew that we might be shut down. And we tried to protect what we had, but the shutdown came faster that what we had realized. We never thought that that would happen so abruptly. And everything was confiscated. Everything was taken out of the offices — all of the files — all the prints in the shop. No. All taken.

## Did Carr know you all were about to be raided?

He knew. And we also knew that we were very close to the edge of being shut down, because Tesla was encouraged to stop because he was threatening the monetary system.

We realized WE were a great threat to the monetary system. Because what we had to offer required once built, the spaceship and the devices from the neutron that could be put in people's homes to accumulate and use electricity. There would be no more need for the grid system and the meters, and so forth, So it was a great threat. And for the automobile industry, and the petroleum industry.

So we knew, and Carr told us, that we don't know if we can get this going before it's shut down. So there was always a knowing that we had to face.

But he also said, "It's inevitable — it has to happen someday." And maybe we're just a middle man on the second or third, or the 4th shift, but someday it will happen. Because it has to.

Because they'll paint themselves into a corner and won't know how to get out unless they go to the visionaries... go to the inventors... go to the discoverers... go to the garage mechanics who are putting these things together... and humbly seek their knowledge and wisdom, because that's where all good things happen in the imagination of the little guys.

# 17
## Where is the OTC-X1 today?

Well, I was told, here — we're in Prescott Arizona right now — and just a mile or two away from here is the Yellowpike College — and I came out here to build a replication of the spaceship under ground — and I designed and build it the same way we built the craft, being very careful to follow natural law.

And they — people who got wind of it came and were very impressed, and they had people coming out and filming, and so forth, and they asked me to speak at Yellowpike College, how I could build a magnificent habitat that was running totally on free and abundant energy, for $5,000 dollars, that was appraised and valued at in those days at $165,000 dollars.

So they asked my to speak — and I did — and a very grainy video still exists — it's somewhere in the archives of the whole episode and the house.

But while I was up at Yavapai College, a professor — this was when the Freedom of Information Act was in progress — and he gave a lecture there and I happened to be in the lecture hall when he was going on about all these things that are happening that we knew nothing about. Now they're known a Black-Ops, but in those days we didn't really have a label for a lot of them, we just felt that they were using the tax payer's money and not letting the people know where the money was going, and so forth.

But he said, one of the things that he noticed — and he didn't even know I was in the audience, but he was mentioning that he was in Los Alamo, New Mexico, and he'd gone down 9 levels underground — 9 elevators, or whatever — down underground and noticed that there were bays filled with different very unusually built devices that looked like they were from out of the future and one of them he way was the OTC-X1, the 45 foot craft that was confiscated.

And the orders of the guards of the place — they were ordered not to talk to any of the observers, but he happened to make friends with one of the guardians, or guards, and he said that they had tried to mount weapons on them, and using a pure magnetic field weapons were obsolete, they can't be used.

So what happened — there was a lot of very tragic and severe accidents because everything they tried would blow up on them. And they had to abandon the idea. They were going to use them as scout vehicles, and so forth.

So I was happy to know that — the location — and I was happy to know that they weren't going to be able to use it, because those machines — the ones we were using — were for humanity, and you had to have the spirit of love and respect, or they wouldn't operate.


## 18
## When you say you had returned to Prescott to build a replica of the craft, can you tell us to what you were referring? To the house, or an OTC craft.?

I had just got done — I had spent some time at UCC Davis — the University of California, Davis — with a Professor Moeller who was designing and building on government grants, vehicles that we're supposed to be levitational, eventually — and they were shaped like a circular foil craft.

But they had 8 small combustion engines around the periphery, and they'd fire them up, and they'd levitate — and the noise was so bad I couldn't even stand it. I couldn't be in there.

So I talked to Moellor and said, you know what, "What about if we considered another source of power, like magnetism?" And he rather abruptly told me, "No, No, this is government funded — this is what they want me to do. This is kind of a show-time for the public to see... see we're working on this."

Then I said, "Do you ever expect to succeed?" And he said, "Not at this rate; we're never going to get more than 100 feet off the ground." So I said, "Well, OK, thank you."

Now he's building floating cars that get maybe 10 or 15 feet off the ground. But the amount of money he was being paid, off government grants, to just make a show that they — *"Well, we're working on it"* — was tremendous. It was hundreds of thousands of dollars. For naught. It wasn't going productively anywhere.


## 19
**That was in David, CA. You previously mentioned coming to Prescott to build replica craft. We're you referring to the house, or the model craft?**

Well, this was the intent. Um, I wanted to become isolated — far enough away from the main stream, so that I wouldn't be bothered or "looky-lood", and so forth, because some of the things we were experimenting with did have things to do with light and vibrations and some sound, and so forth.

So we felt our way over to Prescott Arizona, and actually, about 18 miles north of Prescott was a place called Chino Valley, and we found a very remote piece of land, and we had to have the land invited us because this was part of sacred geometry. You have to be at the right place. You had to start it with a solid foundation or it won't work.

The land invited us through a whole series of very beautiful coincidences. We got the land, and the idea was to build an underground house — and from the levitational point, if we could — but the minute we even thought about levitation, the planning and zoning said there's no such thing. Don't — you're going to have to anchor this, and so forth. So we didn't go any further with that idea. We said, OK.

So we did have the craft [the house] — the replica— it's still there today. We were over there just the other day, I think. You were with us. It's still intact. It's still functional. It's been remodeled. We started... once we got it built to a point I could put up a work bench, I started building small experimental models that I wanted to.

We had 12 acres, 12-1/2 acres and I had a meadow down below, and I wanted to levitate these drafts around, so that — just for tests — to see if we could do it. Well, I didn't have enough of the knowledge, I didn't have enough of what I was doing, so I was taking quite a bit of time.

But we attracted from the neighbors — because we were lighting up the whole place and we had no wires, no telephone poles. We had plenty of water. We were using air wells for water. We were using ham radios for telephones, and solar panels, and we had television. We had computers, small computers., 16 KV, but that's what it was in those days. But everything was running. And we lit up the whole top of the mountain, and our neighbors were using candlelight, and lanterns, and flashlights — we became a very curious friend to them.

So they had people coming out — well, they would report them to somebody, and they would come out and kind of see, *"What are you up to here",* and so forth. *"Are you following the codes and rules?"* Because the septic system that we put in — I didn't want to go with that — I had plans from Sweden about a recyclable system that — completely pure and reusable, and the methane gas and everything could be used. And I went to planning and zoning, and they said, "Forget it. You've gotta dig a whole septic system — put in leach lines — I had the adhere to something I didn't want to.

But we became apprehensive because they were then starting to fly — see I was being watched ever since I left Carr — they said, "We're going to be monitoring you forever, and you're not to get — even try this, or get together with anyone." So then when we started noticing these black helicopters with no marking on them, and they's come right over with big cameras on them, and I could see the guy with the cameras, just filming our land and trying to film inside of what was the underground house — we felt this isn't going to work — because they're going to come in and take our land away — or whatever.

So we put it into prayer and turned it over to natural law, that what the next step would be, would come to us.

We didn't even try to plan anything, and a gentleman by the name of Walter Baumgartner, who wrote the magazine, "Energy Unlimited", he wrote volumes after volumes of the type of work we were doing — and he even had Tesla's work in there — he came over and said, I've heard about you, and I don't really know how he had heard about us, because we were trying to keep things kind of quiet, and I said we're having difficulty here because we've been kinda spotted and we've been told if we...

And he said don't worry about that, I've got a place in Magdelena in New Mexico, totally on an Indian reservation, totally isolated, and I've got all of the facilities, machine shop, tool and die, to help you Will you work with me? And I said sure. So we gave him everything we have, so that if anybody come in with an idea of finding something that they could cause a problem for us, we gave it to Walter. He took it back over to New Mexico, and very delightfully, we were commuting back and forth to New Mexico, and he was building small replicas and large replicas over there. I do still have some of those plans and things somewhere.

But he got so far — and the unusual thing that most people don't understand, is that this isn't a conventional nuts and bolts linear craft. It's not like anything... it's now like an automobile. You can't put in a key, turn it on, and run it like an automobile. It's run by consciousness. It's run by synogizing with the craft itself. And when you're working with it you have to be very very humble and close to nature because every detail has be honored for what it is. And he built it, And he was successful in levitating it off his platform in inches, and it took him a tremendous amount of work to get it that far, and he became very very frustrated and kept coming back and forth, and we kept staying over.

And I said, "You really have to have the spirit. You really have to love this. This has to be alive. Everything, when you understand it, is energy. But energy is live. It's all alive consciousness. And in order to effectively get off the ground, with the type of device, you have to get very very close to the consciousness of what you're working with. You have to love every part and love the way it sets and feel how it sets, and if it doesn't feel right, tear it up and start over. Because it won't work unless you feel your way through it. If you don't think your way through it, you'll never get anywhere.

Because thought is very limited. It just doesn't care much about anything outside of its jurisdiction which is a 3rd dimensional reality, 3rd dimensional body. Anything foreign to that is a truest, so it ignores, or sets up a defense, or even becomes very aggressive, if there is anything outside of its jurisdiction.

And he had those blocks that he couldn't get out of the way and allow his spirit of loving what he was doing, even though he was an engineer and he enjoyed the challenge, he eventually — we lost contact because we had to do some other things, and eventually he had to cease his operation. And I don't know, I never talked to him in the final days, or whatever, so...

And I heard, he's gone to Canada, and was trying to affect building them in Canada with somebody. But that's the last I ever heard of him.

## 20

**One of the drawings show a Van de Graaff generator connected to the 'Central Power Unit' (Central Accumulator / Central Utron)**

Oh, well, that was abandoned at one tine. I mean, we were — a lot of these were, some of them anyway, are earlier drawings and, and some of the later drawings aren't ever here. The accumulator, once aligned with the natural flows of energy — the magnetic fields of energy that are flowing all around us are accumulated electricity, if you will, or energy, not really electricity until it is accumulated, and then distributed to, as opposed to a generator which generates electricity. And this is what we live in now, generated electricity which is sold through telephone poles and wires, and eventually a meter and become very very expensive, causing a displacement sometimes of 78-80% of a persons income must to afford the energy bills. So we wanted to replace them with this simple Utron, which is an accumulator, and once put in the craft — an RV in the sky — you can take off, and you can stay a few inches, or a few miles above the surface. You can never have and accident, magnetic fields are accident free. There's no noise, no pollution, there's no G-forces to experience. It's a different mousetrap. It's completely different.

## 21

**This picture of a model OYC-X1 shows a wire sticking out of the top of it. Is that an aerial or an antenna?**

Yeah. Aerials are essential. When we were experimenting, we had to have the grounding which was the earth. And we had an antenna — an aerial — which you could say, the positive and negative — understanding of how energy flows, like the Mobius twist — everything is in vortexual motion, it never really begins and ends.

And in order to tap it, at a point when it's going to be beneficial, you have to have an antenna, and you have to have a grounding, per se, in our experiments, and later on we found we didn't need that because there were other way of obtaining the ground. We used the magnetic field around the earth and we, you might say, 'tweaked' it, so we had a grounding, and it wasn't necessary to ground ourself.

## 22
## Does this technology represent a threat to monetary or other systems?

There is no threat to the System as far as people wanting the horse and buggy.

The Amish still have the horse and buggy. The Quakers have the horse and buggy. The car lovers, enthusiasts, classic cars are still used, Corvets are still used, Maseratis are still used, they don't have to give any of that up. We just want to add to the system.

We want to... like they add airplanes, we want to add the vehicles — circular foil craft that would be both habitation and transportation. And they could be as large as you wanted. You could have farms, you could have waterfalls, you could have lakes, you could have anything on those. It doesn't matter. The technology I am speaking of doesn't sound very familiar because its' ether, in essence, and it has to do with the creative powers of the mind and once you realize how creative we really are, you can do anything. There's nothing that's impossible to do. And we've done it. We've been there and done it, and now we want to present this over and over again until they allow us the privilege of showing the world what we have.

## 23
## Is the Central Accumulator filled with an electrolyte or mercury? Is it solid or hollow?

Well we experimented with both, and it took Carr years to finally — you know in many many experiments it took lots of time — lots of money to finally arrive at a very hollow double tetrahedron — shaped like a diamond — or 2 ice-cream cones, one on top of the other at the vase. And this being — because it contained all the geometric figurations of tectonic solids which — you know there's the square, the circle, etc, etc, the pyramid, all in one object which is now being called the accumulator.

And once you take that and you understand through sacred geometry theories of space, the magnetic fields. Everything has a magnetic field around it, through it, and in it, and once you understand that, then you can accumulate energy, and no batteries are necessary. The accumulation and distributor are all one and it goes on forever, there's no need for batteries, except you're actually piggy backing nature — energy itself.

There is nothing but energy. And you're piggy backing that and using it as such. You're grabbing on and it would be like grabbing on to a street car and going, where your destination might be, a mile away, and then letting go. It's the same thing with energy. It's everywhere. So you have only to grab onto it, to use it. And it's inexhaustible. It goes on forever.

## 24
### What materials were used for the various parts of the OTD-X1, such as insulators, equatorial field correlator, and the large flat capacitors?

I really don't know, because the materials in those days were very aniquated compared with what we have today, and to say silicone, or poly-anything, would be a mistake, because I don't know what it was. I'm not sure.

## 25
### OTC bulk-ordered large sheets of T6 Aluminum for the outer skin of the craft?

Yah. That's right. Yah. I don't know where they are now, in aluminum, but in those days, that was the best he could get, and that's what we wanted. Um. The other parts of the craft, some of them were conventional parts. I machined come of the little parts and stuff. I don't really don't remember too much about the details of anything.

## 26
### The Central Accumulator is a large hollow aluminum utron. The outer utrons on the rotating central assembly are solid aluminum and don't spin?

Yeah. True. The outer ones differ in the sense that they were solid, but they were suspended on both tips like the diamond, the double tip tetrahedron so that they could turn. So if the natural flows of energy caused them to turn, so what, we didn't care, and it may could have been used to its advantage, if we understood it, but it wasn't necessary. We just had them suspended, and they were going through horseshoe magnets.

## 27
## OTC has referred to the Utron as a coil. Is it?

The answer to that has to be multi because it's not a single answer. Because a generator is a motor, and a motor is a generator. It just depends on how you're using it. On your windings, on your understanding of what a generator does, and what a motor does, and so forth.

The Accumulator is and was capable of doing many many things, you know. It wasn't just to operate a space ship, per se, and the initial idea was to make up small prototype boxes, if you will, and call them Black Boxes, with an antenna on it, and give everyone a box for their home, eventually, that they could... They'd have to take the responsibility of dealing with the power company, or whatever. We would at least as far as our finances would go, give them out to as many people as we could, to have them be our advertising agency, and PR work, so they'd say, *"Yes, I've had one those things and its been working fine"*, and so forth, but we never got that far.

The questions is, *the answer to the question* — the Accumulator was not built just for the spaceship — it was built to tie-in with the laws of nature — to the oneness of it all.

That is — the oneness is the inexhaustible energy that's always flowing. And that could be used anywhere. You could put it... if you had a small accumulator, you could put it in perfect alignment... you could put it on anything. Any conventional thing could now be properly adapted. The plug could be pulled from the wall, and you could operate it.

## 28
## What are your recollections about the 45 foot draft's main shaft or spindle?

Yeah. The spindle. I don't really... the spindle — it must have, I don't recall the spindle ever going through the accumulator, but maybe it did — I never say the inside of it. I knew that it was hollow — there was nothing in there —and I would think that the infusion of anything in there would disrupt the sacred geometry — unless it was designed to work that way. But I don't know.

## 29
## Some say they don't have technical or mechanical skills; but want to help. How can they help?

The most effective way they can be helping in this entire agenda — in this entire intention is to be who and what they really are. To not hold anything in pretention any more and not to have to say 'Yes' to something they mean 'No' to, or 'No' to something they mean 'Yes' to. To align their heart with their mouth, with their brain, and say what they really feel, and do what they really feel.

This will affect the change necessary, not only in them, but in the collective consciousness which is all of us, and If we can do this collectively together. If we can get the idea of standing our ground, and with the threat of being fired, with the threat of loosing your home, or car — stand your ground — because you have the power of the multi-verse with you. You don't have the threats and worry and fears, and doubts of the 3rd dimension any longer, if you don't want it.

We bring these things on because we give our power away because of fear and doubts. And we have to take our power back by realizing simply, I'M ME! I can't help it. I love it. Just love yourself. Love what you're doing.

And brace yourself. There are tests that you have to go through. In some cases, called the ring, pass not. In Buddhism there are many and many initiations that you have to go through in life before you can achieve a state of knowing, of being still and peaceful inside. And then along comes the joy, and then eventually, you'll find bliss in realizing that everywhere you go is heaven. Everything you do.

If you want something to change, you can have a contract — because you can create it. Do you want disturbance? You can create it. It's in control of you — nobody else. So the best thing in the way of helping would be to raise the consciousness. Of course, we have people in the field now spending long hours, day and night, like Walter's feeling this episode right now. He's coming here with just his honor his dedication, his devotion to seeing that these words and ideas are getting out to the people. No pay. Nobody really sponsoring or taking care of his needs of course.

And there are people in our Pods — we call them Pods — there just small groups of people — there all over the world — in Australia, New Zeland, and everywhere, working on a shoestring, so if you happen to have money that you really don't know what to do with and you'd like to make a better world it is, that's up to you, but we're not asking for it, but I'm covering all bases, in the answer to the questions. Whatever you feel dedication and motivation to what we're trying to do will be welcome, and used effectively to affect a change.

30

## As a knight yourself, can you speak to the difference between the Knights and the Freemasons; and to how secrecy is reconciled with Natural Law?

First of all, there are 3 well known stages of knighthood. One is the Knights of Malta — One is the Knights of St. John, which are affiliated with the knights of Malta, and then there's the Knights Templar. That's the 3 well known knights.

There are knights everywhere in the world on different levels, who just go 'round affecting changes, and helping people — not asking for anything in return. That's what true knighthood is. It doesn't want to do anything but protect the rights of others to do whatever in life they want.

The changes now in knighthood being observed on a very large and deeply rooted scale, that being that they were and are to this day an organization ... They are a corporate organization that has rules and regulations based on 3rd dimensional realities, and contracts, etc, etc, etc, in a sense.

It's not as drastic as the contracts we have with General Motors, or T.G. &E., or whatever, but it is still a Monarchy where their being told, you must adhere to this and that — like a religion — its similar to a religion, although it's much deeper because a knight is dedicated, and you have to be dedicated — you're a knight, before your a knight — you have to have it in your heart and soul — to want to be a guardian of people's rights to pursue their life in any way they want. No restrictions. No, nothing.

And the organizations have restrictions. Where you really have to do this, and so forth. No, they're dissolving their organization. And many of us... Marcia and I were both knights in the Knights Templar. We pulled away from the structure and have become what's called rogue knights. We still now are free to be silent knights. We can go out and attempt changes. Just what we're doing. Quietly. At no charge.

We put out lectures all over the place, that are at our expense, and so forth. Because we enjoy doing it, and that joy brings this great satisfaction in the people realizing what we are taking about, and waking up, and letting us know, *"Wow! I never knew that!"* And things like that. That's where we get our energy.

So, the Knights of Malta... we have friends who are Knights of Malta, and the knights of St. John, and the Knights Holpitallers. We have one in one of our Pods. He's located in Australia. It's call "Path of Divine Restoration." And it's run by knights that are Rogue Knights. And we are doing, and being, effective in changing the realities of what's been going on in the way of tyranny and the way of servitude, and so forth.

They're very actively alive and they have pulled away from structure, so they're now all Rogue Knights. There are knights in Canada, the same thing. They're becoming Rogue Knights, pulling away from the structure — which frees us to be who and what we are — to be on our own free will — to be responsible for our own actions.

We don't have to adhere to anybody, except our Source, which is God. And when we're in touch with that Source, we're being led where we can help and affect a positive and constructive change in reality. And so on it goes. The structures now... all corporate structures now are feeling the effects of what's happening. Everything is in a flux of change, and they are disintegrating, because they were built on grains of sand instead of a solid foundation.

The first thing... the first choice... the first decisions that have been made were based on a profit margin... in stead of on the idea of making a better life for everybody, regardless of the money... and when you base things on a profit margin, and you become very very sloppy in your theory, you don't particularly care where or how you can fluctuate the monetary system up on down according to what the traffic will bear, and this puts people at a very awkward situation.

But those that were built on the premiss of a profit return are now starting to crumble, and they will all fall apart — they can't hold up. Those that have been based on a humanitarian interest — both human, and animal, and plant life, and caring for those things, will flourish, and become, now, the new and different paradigm that will return this earth to what it was always meant to be — a beautiful garden.

And we can visit it off the ground once and awhile. So, that's the Knights. We're very humbly honored to be Knights, and it isn't something that happens over nights, it takes a lot of dedication, a lot of knowing that everyone's life must become more important than your own, at all times. No matter where you go, or what you do. That's where we stand.

## 31
### Is there anything else you'd like to add?

It would be appropriate to say that these things, and more, are now waiting in part around the earth. It's global in scope. And people have only to raise their scopes — their horizons, or their consciousness, to start accepting things that they have been programed in the past not to accept. As a sense of reprogramming saying and doing, and realizing.

Everything is a part of the One. And the One is good. There is nothing that is not good.

We have created the idea of fear and doubt, and this is our creation — it is not nature's. And if we refuse to no longer use those, our consciousness will raise all of itself to a higher degree, and higher understanding, and a deeper love and respect for nature and all of the laws that go with it.

It has only to be realized that we are energy. We are materializing things with our mind, and this density call the 3rd dimension is being created in what could equate to nano seconds. It's so fast that the brain cannot even conceive of the conception of manifestation.

But the minute we look at something, we're creating. The minute we think of something, we're creating. We're infinite creators. And we go on an on forever. And if we would create good and wonderful things, and be productive in what we think about, what we feel, what we know, and look at things with compassions, and love instead of fears, and doubts, and ideas of profit margins, we could have the realization that we have never left heaven; that it's always been here, we just lost sight of it.

And in raising our consciousness you'll find joy in what you're doing. Joy in what you realize and eventually feel every where you go is heaven, because it's not outside yourself — It's inside, and once you find that, you never want to go any place else.

I guess that would... I world like to see everybody ponder that for awhile, and realize I had my... part of my body... my finger... looked at on a microscope... and all I saw was energy.

So when you get to realizing that you're energy, and a magnificent created being, and you don't have to succumb to servitude, you don't have to toil anywhere under somebody's dictatorship for the way and means to take care of yourself and you family, you can stand your ground and realize that you are part of an infinite creation, and a creator — which is all of us — and collectively together there's nothing that can stop us.

The thing that has separated us — and continues to separate us is the programming that says we're wrong, we're bad, we can't, we should, we shouldn't. And those things now, because we're entering a new paradigm, have to be abandoned, and the new ideas, simple ones, like, I can do anything, everything is good, everything is beautiful, and I won't accept anything else.

Once you accept or adopt that attitude, you'll find your whole world changing, dramatically, beautifully, differently. And you don't have to worry because your needs will come to you. Don't try. Just stay out of the way, and keep that consciousness alive that everything is good, everything is OK.

We're living in the Now.

The mistakes that the 3rd dimensional make, is taking the past and trying to put it into the future.

A brief example of that is, you have a problem coming up, and it's looking like the same problem we had before, so I'll use the same technique as the past. Well, the problem still comes back. So apparently it was never solved. So it keeps coming back, stronger some times, because that's the way nature kind of gives wake up calls, because you're taking the past understanding, and trying to solve a future situation, and it will never work. It will just be in a squirrel cage forever.

You have to be still, and live in the Now, and stay in the Now with the faith and the trust, and the discipline — because it is not easy — you want of give in to temptation. You want of give in to fears and doubts, and panic — but don't go there — just stand your ground, and stay there long enough, and you'll find that you'll never be let down by the laws of nature, because it's always with you, and if you keep talking the trash out of your thought patterns — thought vs. your feelings — which is your intuition — if you keep turning those out and replacing them with your feeling of love and compassion, and so forth, you'll find your whole world — instantly sometimes

— will change. And everything you need will come to you, when you need it. — not necessarily when you want it — but it will be there when you need it.

When something is really necessary for your growth and understanding, it will be there, and it can't fail because you're the creator of your own reality and your own destiny and you are creating the need for something to take care of yourself, your family, whatever, and it will come. In an different way, perhaps, but it will come.

# 14
# Many Events to Unfold

Many events are set to unfold which will clearly demonstrate that a large number of people are successfully releasing themselves from the bondage of fear and from unloving attitudes encouraged by fear. Humanity cannot and will not slip back into the old ways that led to repression, conflict, and war. We're on an evolutionary path, motivated by love and compassion for all beings, and we will not be turned back or diverted from that path.

Needless to say, there are still many who favor the old ways and expect to prevent humanity from making changes that are essential for our spiritual evolution and growth, but they will not succeed. Their power and influence are rapidly declining, much to their surprise and dismay. They are in a state of confusion, and are unable, and will remain unable, to resolve their many points of disagreement, making it impossible for them to continue holding the reins of power with which they have controlled so many for so long. Their disarray will lead to their collapse, and people of honor and integrity will move into positions of leadership to contain and resolve the many damaging crises that have been causing so much suffering worldwide.

Attitudes are changing all over the world as the realization dawns on more and more people that the old ways of threat, bluff, confrontation, and dishonesty in attempting to repair and heal damaged relationships are totally ineffective. This has always been the case, but now people are finally saying "There *must* be a better way!", and they are discovering that indeed there is – open honest discussion which includes full disclosure of the needs and desires of all parties with absolutely no hidden agendas. When discussions take place in this kind of atmosphere, with all working compassionately and empathetically together, there are truly no insoluble problems anywhere in the world.

A time of great opportunity for all on earth to move into life-styles with which they agree is approaching, because the loving attitudes that so many people have now adopted make it inconceivable that anyone would be coerced into doing work which is unsuited to them. Abilities, skills, talents, and competencies will no longer be judged competitively, so there will be no authoritarian hierarchical organizations where respect and honor is paid to positions held. Instead *all* will be equally honored and respected, regardless of job title or qualification, solely for who they are – beloved children of God, and sisters and brothers to each other.

The divisiveness of the old ways is ending. Everyone on the planet deserve loving acceptance and respect without judgment or evaluation of their skills, their position in the community or organization, their race, their culture, their beliefs, or their life-styles. And because people are more and more releasing old unloving and divisive attitudes this will occur spontaneously, as all sense of threat or disapproval disappears, and is replaced by fascination and a desire to understand.

These changes will inspire creative ventures in which all who feel attracted to them can take part. Everyone will be encouraged and assisted in developing skills and competencies in any field that appeals to them, and there will be mentors available to guide and inspire them to uncover and develop talents and skills which previously had lain hidden within them.

A new age full of amazing and creative concepts abd ideas is dawning, in which everyone will be able to effectuveky participate, giving rise to unprecedented happiness and satisfaction everywhere.

Hope Deferred

# 3
# Echoes of the Past

More than 125 years ago, in the late 1880's, trade journals in the electrical sciences were predicting free electricity and free energy in the near future. Incredible discoveries about the nature of electricity were becoming common place. Nikola Tesla was demonstrating *"wireless lighting"* and other wonders associated with high frequency currents. There was an excitement about the future like never before!

The Victorian Age was giving way to something totally new. Within 20 years, there would be automobiles, airplanes, movies, recorded music, telephones, radio, and practical cameras.

For the first time in history, common people were encouraged to envision a utopian future filled with abundant modern transportation and communication, as well as jobs, housing and food for everyone. Disease would be conquered, and so would poverty. Life was getting better, and this time, everyone was going to get a piece of the pie.

So, what happened? In the midst of this technological explosion, where did the energy breakthroughs go? Was all of this excitement about free energy, which happened just before the beginning of the last century, just wishful thinking — that "real science" eventually disproved?

## Current State of Technology

The answer to that question is No. In fact, the opposite is true. Spectacular energy technologies were developed right along with the other breakthroughs. Since that time, multiple methods for producing vast amounts of energy at extremely low cost have been developed behind the scenes. However, none of these technologies have made it to the consumer market as an article of commerce.

Exactly why this is true will be discussed, shortly.

But first, here is a short list of free energy technologies that we sould be currently aware of, that are proven beyond all reasonable doubt. The common feature connecting all of these discoveries, is that they use a small amount of *one form of energy* to control or release a large amount of *a different kind of energy.* Many of them tap the underlying **Æther field** in some way; a source of energy conveniently and intentionally ignored by modern science.

### 1) Radiant Energy.

Nikola Tesla's magnifying transmitter, T. Henry Moray's radiant energy device, Edwin Gray's EMA motor, and Paul Baumann's Testatika machine all run on radiant energy. This natural energy form can be gathered directly from the environment (mistakenly called "static" electricity) or extracted from ordinary electricity by the method called fractionation. Radiant energy can perform the same wonders as ordinary electricity, at less than 1% of the cost. It does not behave exactly like electricity, however, which has contributed to the scientific community's misunderstanding of it. The Methernitha Community in Switzerland currently has 5 or 6 working models of fuelless, self-running devices that tap this energy today.

## 2) Permanent Magnets.

Dr. Robert Adams (NZ) has developed astounding designs of electric motors, generators and heaters that run on permanent magnets. One such device draws 100 watts of electricity from the source, generates 100 watts to recharge the source, and produces over 140 BTU's of heat in two minutes! Dr. Tom Bearden (USA) has two working models of a permanent magnet powered electrical transformer. It uses a 6-watt electrical input to control the path of a magnetic field coming out of a permanent magnet. By channeling the magnetic field, first to one output coil and then a second output coil, and by doing this repeatedly and rapidly in a "ping-pong" fashion, the device can produce a 96-watt electrical output with no moving parts. Bearden calls his device a Motionless Electromagnetic Generator, or MEG. Jean-Louis Naudin has duplicated Bearden's device in France. The principles for this type of device were first disclosed by Frank Richardson (USA) in 1978. Troy Reed (USA) has working models of a special magnetized fan that heats up as it spins. It takes exactly the same amount of energy to spin the fan whether it is generating heat or not. Beyond these developments, multiple inventors have identified working mechanisms that produce motor torque from permanent magnets alone.

### 3) Mechanical Heaters.

There are two classes of machines that transform a small amount of mechanical energy into a large amount of heat. The best of these purely mechanical designs are the rotating cylinder systems designed by Frenette (USA) and Perkins (USA). In these machines, one cylinder is rotated within another cylinder with about an eighth of an inch of clearance between them. The space between the cylinders is filled with a liquid such as water or oil, and it is this "working fluid" that heats up as the inner cylinder spins. Another method uses magnets mounted on a wheel to produce large eddy currents in a plate of aluminum, causing the aluminum to heat up rapidly. These magnetic heaters have been demonstrated by Muller (Canada), Adams (NZ) and Reed (USA). All of these systems can produce ten times more heat than standard methods using the same energy input.

### 4) Super-Efficient Electrolysis.

Water can be broken into hydrogen and oxygen using electricity. Standard chemistry books claim that this process requires more energy than can be recovered when the gases are recombined. This is true only under the worst case scenario. When water is hit with its own molecular resonant frequency, using a system developed by Stan Meyers (USA) and again recently by Xogen Power, Inc., it collapses into hydrogen and oxygen gas with very little electrical input. Also, using different electrolytes (additives that make the water conduct electricity better) changes the efficiency of the process dramatically. It is also known that certain geometric structures and surface textures work better than others do. The implication is that unlimited amounts of hydrogen fuel can be made to drive engines (like in your car) for the cost of water. Even more amazing is the fact that a special metal alloy was patented by Freedman (USA) in 1957 that spontaneously breaks water into hydrogen and oxygen with no outside electrical input and without causing any chemical changes in the metal itself. This means that this special metal alloy can make hydrogen from water for free, forever.

### 5) Implosion/Vortex.

All major industrial engines use the release of heat to cause expansion and pressure to produce work, like in your car engine. Nature uses the opposite process of cooling to cause suction and vacuum to produce work, like in a tornado. Viktor Schauberger (Austria) was the first to build working models of implosion engines in the 1930's and 1940's. Since that time, Callum Coats has published extensively on Schauberger's work in his book Living Energies and subsequently, a number of researchers have built working models of implosion turbine engines. These are fuelless engines that produce mechanical work from energy accessed from a vacuum. There are also much simpler designs that use vortex motions to tap a combination of gravity and centrifugal force to produce a continuous motion in fluids.

### 6) Cold Fusion.

In March 1989, two chemists from the University of Utah (USA) announced that they had produced atomic fusion reactions in a simple tabletop device. The claims were "debunked" within six months and the public lost interest. Nevertheless, cold fusion is very real. Not only has excess heat production been repeatedly documented, but also low energy atomic element transmutation has been catalogued, involving dozens of different reactions! This technology definitely can produce low cost energy and scores of other important industrial processes.

### 7) Solar Assisted Heat Pumps.

The refrigerator in your kitchen is the only free energy machine you currently own. It's an electrically operated heat pump. It uses one amount of energy (electricity) to move three amounts of energy (heat). This gives it a co-efficient of performance (COP) of about 3. Your refrigerator uses one amount of electricity to pump three amounts of heat from the inside of the refrigerator to the outside of the refrigerator. This is its typical use, but it is the worst possible way to use the technology. Here's why. A heat pump pumps heat from the source of heat to the "sink" or place that absorbs the heat. The source of heat should obviously be hot and the sink for heat should obviously be cold for this process to work the best. In your refrigerator, it's exactly the opposite. The source of heat is inside the box, which is cold, and the sink for heat is the room temperature air of your kitchen, which is warmer than the source. This is why the COP remains low for your kitchen refrigerator. But this is not true for all heat pumps. COP's of 8 to 10 are easily attained with solar assisted heat pumps. In such a device, a heat pump draws heat from a solar collector and dumps the heat into a large underground absorber, which remains at 55° F, and mechanical energy is extracted in the transfer. This process is equivalent to a steam engine that extracts mechanical energy between the boiler and the condenser, except that it uses a fluid that boils at a much lower temperature than water. One such system that was tested in the 1970's produced 350 hp, measured on a Dynamometer, in a specially designed engine from just 100-sq. ft. of solar collector. (This is not the system promoted by Dennis Lee.) The amount of energy it took to run the compressor (input) was less than 20 hp, so this system produced more than 17 times more energy than it took to keep it going! It could power a small neighborhood from the roof of a hot tub gazebo, using exactly the same technology that keeps the food cold in your kitchen. Currently, there is an industrial scale heat pump system just north of Kona, Hawaii that generates electricity from temperature differences in ocean water.

There are dozens of other systems that have not been mentioned, many of them are as viable and well tested as the ones recounted. But this short list is sufficient to make the point: free energy technology is here, NOW. It offers the world pollution-free, energy abundance for everyone, everywhere.

It is now possible to stop the production of greenhouse gases and shut down all of the nuclear power plants. We can now desalinate unlimited amounts of seawater at an affordable price, and bring adequate fresh water to even the most remote habitats. Transportation costs and the production costs for just about everything can drop dramatically. Food can even be grown in heated greenhouses in the winter, anywhere.

All of these wonderful benefits that can make life on this planet so much easier and better for everyone have been needlessly postponed for decades. Why? Whose purposes are served by this postponement?

### Four Invisible Forces

There are four gigantic forces that have worked together to create this situation. To say that there is and has been a conspiracy to suppress this technology only leads to a superficial understanding of the world, and it places the blame for this completely outside of ourselves. Our willingness to remain ignorant and actionless in the face of this situation has always been interpreted by two of these forces as **implied consent**. So, besides a non-demanding public, what are the other forces that are impeding the availability of free energy technology?

In the United States, and in most other countries around the world, there is a **money monopoly** in place. I am free to earn as much money as I want, but I will only be paid in Federal Reserve Notes. There is nothing I can do to be paid in Gold Certificates, or some other form of money. This **money monopoly** is solely in the hands of a small number of private stock banks, and these banks are owned by the wealthiest families in the world. Their plan is to eventually control 100% of all of the capital resources of the world, and thereby control everyone's life through the availability (or non-availability) of all services and goods.

An independent source of wealth (free energy device) in the hands of each and every person in the world, ruins the plans of the wealthiest families for world domination, permanently. Why this is true is easy to see. Currently, a nation's economy can be either slowed down or sped up by the raising or lowering of interest rates. But if an independent source of capital (energy) were present in the economy, and any business or person could raise more capital *without borrowing it from a bank*, this centralized throttling action on interest rates would simply not have the same effect.

Free energy technology changes the value of money.

The wealthiest families and the issuers of credit do not want any competition. It's that simple. They want to maintain their **current monopoly control** of the money supply. For them, free energy technology is not just something to suppress, it must be permanently forbidden!

So, the wealthiest families and their central banking institutions are the first force operating to postpone the public availability of free energy technology. Their motivations are **the imagined divine right to rule**, greed, and their insatiable need to control almost everything except themselves. The weapons they have used to enforce this postponement include intimidation, "expert" debunkers, buying and shelving of technology, murder and attempted murder of the inventors, character assassination, arson, and a wide variety of financial incentives and disincentives to manipulate possible supporters. They have also promoted the general acceptance of a scientific theory that states that free energy is impossible (laws of thermodynamics).

The second force operating to postpone the public availability of free energy technology is national governments. The problem here is not so much related to competition in the printing of currency, but in **the maintenance of national security**. The fact is, the world out there is a jungle, and humans can be counted upon to be very cruel, dishonest, and sneaky. It is government's job to provide for the common defense. For this, police powers are delegated by the executive branch of government to enforce "the rule of law." Most of us who consent to the rule of law do so because we believe it is the right thing to do, for our own benefit. There are always a few individuals, however, that believe that their own benefit is best served by behavior that does not voluntarily conform to the generally agreed upon social order. These people choose to operate outside of the rule of law and are considered outlaws, criminals, subversives, traitors, revolutionaries, or terrorists.

National governments have discovered, by trial and error, that the only foreign policy that really works, over time, is a policy called "tit for tat." What this means to you and me is, that **governments treat each other the way they are being treated**. There is a constant jockeying for position and influence in world affairs, and the strongest party wins! In economics, it's the Golden Rule, which states: "The one with the gold makes the rules."

So it is with politics also, but its appearance is more Darwinian. It's simply survival of the fittest. In politics, however, the fittest has come to mean the strongest party who is also willing to fight the dirtiest. Absolutely every means available is used to maintain an advantage over the adversary, and **everyone else is the adversary whether considered friend or foe**. This includes outrageous psychological posturing, lying, cheating, spying, stealing, assassination of world leaders, proxy wars, alliances and shifting alliances, treaties, foreign aid, and the presence of military forces wherever possible.

Like it or not, **this is the psychological and actual arena national governments operate in**. No national government will do anything that simply gives an adversary an advantage for free. It's national suicide. An activity by any individual, inside or outside the country, that is interpreted as giving an adversary an edge or advantage will be deemed a threat to "national security."

**Free energy technology is a national government's worst nightmare!**

Openly acknowledged, free energy technology sparks an unlimited arms race by all governments in a final attempt to gain absolute advantage and domination. Think about it. Do you think Japan will not feel intimidated if China gets free energy? Do you think Israel will sit by quietly as Iraq acquires free energy? Do you think India will allow Pakistan to develop free energy? Do you think the USA would not try to stop Osama bin Ladens from getting free energy?

Unlimited energy available to the current state of affairs on this planet leads to an inevitable reshuffling of the balance of power. This could become an all-out war to prevent "the other" from having the advantage of un-limited wealth and power. Everybody will want it, and at the same time, want to prevent everyone else from getting it.

So, **national governments are the second force** operating to postpone the public availability of free energy technology. Their motivations are "self-preservation." This self-preservation operates on three levels.

**First,** by not giving undue advantage to an external enemy.

**Second,** by preventing individualized action capable of effectively challenging official police powers (anarchy) within the country.

**Third,** by preserving income streams derived from taxing energy sources currently in use.

Their weapons include the preventing of the issuance of patents based on national security grounds, the legal and illegal harassment of inventors with criminal charges, tax audits, threats, phone taps, arrest, arson, theft of property during shipment, and a host of other intimidations which make the business of building and marketing a free energy machine practically impossible.

The third force operating to postpone the public availability of free energy technology consists of the group of **deluded inventors and out right charlatans and con men**. On the periphery of the extraordinary scientific breakthroughs that constitute the real free energy technologies, lies a shadow world of unexplained anomalies, marginal inventions and unscrupulous promoters. The first two forces have constantly used the media to promote the worst examples of this group, to distract the public's attention and to discredit the real breakthroughs by associating them with the obvious frauds.

Over the last hundred years, dozens of stories have surfaced about unusual inventions. Some of these ideas have so captivated the public's imagination that a mythology about these systems continues to this day. Names like Keely, Hubbard, Coler, and Henderschott immediately come to mind. There may be real technologies behind these names, but there simply isn't enough technical data available in the public domain to make a determination. These names remain associated with a free energy mythology, however, and are sited by debunkers as examples of fraud.

The idea of free energy taps very deeply into the human subconscious mind. A few inventors with marginal technologies that demonstrate useful anomalies have mistakenly exaggerated the importance of their inventions. Some of these inventors also have mistakenly exaggerated the importance of themselves for having invented it. A combination of **"gold fever"** and/or **a messiah complex** appears, wholly distorting any future contribution they may make.

While the research thread they are following may hold great promise, these deluded inventors begin to trade enthusiasm for facts, and the value of the scientific work from that point on suffers greatly. There is a powerful, yet subtle seduction that can warp a personality if they believe that the world rests on their shoulders or that they are the world's savior. Strange things also happen to people when they think they are about to become extremely rich.

**It takes a tremendous spiritual discipline to remain objective and humble in the presence of a working free energy machine.**

Many inventors' psyches become unstable just believing they have a free energy machine. As the quality of the science deteriorates, some inventors also develop a persecution complex that makes them very defensive and unapproachable. This process precludes them from ever really developing a free energy machine, and fuels the fraud mythologies tremendously.

Then there are the out right con men. In the last 15 years, there is one person in the USA who has raised the free energy con to a professional art. He has raised more than $100,000,000, has been barred from doing business in the state of Washington, has been jailed in California, and he's still at it. He always talks about a variation of one of the real free energy systems, sells people on the idea that they will get one of these systems soon, but ultimately sells them only promotional information which gives no real data about the energy system itself. He has mercilessly preyed upon the Christian community and the patriot community in the USA, and is still going strong.

This man's current scam involves signing up hundreds of thousands of people as locations where he will install a free energy machine. In exchange for letting him put the free energy generator in their home, they will get free electricity for life, and his company will sell the excess energy back to the local utility company. After becoming convinced that they will receive free electricity for life, with no out-front expenses, they gladly buy a video that helps draw their friends into the scam as well. Once you understand the power and motivations of the first two forces we have discussed, its obvious that this person's current business plan cannot be implemented. This one person has probably done more harm to the free energy movement in the USA than any other force, by destroying people's trust in the technology.

So: **The third force** postponing the public availability of free energy technology is **delusion and dishonesty within the movement itself.** The motivations are self-aggrandizement, greed, want of power over others, and a false sense of self-importance. The weapons used are lying, cheating, the "bait and switch" con, self-delusion and arrogance combined with lousy science.

**The fourth force** operating to postpone the public availability of free energy technology is **all of the rest of us.** It may be easy to see how narrow and selfish the motivations of the other forces are, but actually, these motivations are still very much alive in each of us as well. Like the Wealthiest Families, don't we each se-cretly harbor illusions of false superiority, and the want to control others instead of ourselves? Also, wouldn't you "sell out" if the price were high enough, say, take a mil-lion dollars, cash, today? Or like the governments, don't we each want to ensure our own survival? If caught in the middle of a full, burning theater, do you panic and push all of the weaker people out of the way in a mad, scramble for the door? Or like the deluded inventor, don't we trade a comfortable illusion once in a while for an uncomfortable fact? And don't we like to think more of ourselves than others give us credit for? Or don't we still fear the unknown, even if it promises a great reward?

You see, really, all four forces are just different aspects of the same process, operating at different levels in the society. There is really only one force preventing the public availability of free energy technology, and that is the unspiritually motivated behavior of humans.

**Free energy technology is an outward manifestation of divine abundance.**

It is the engine of the economy of an enlightened society where people voluntarily behave in a respectful and civil manner toward each other, and where each member of the society has everything they need, and does not covet what their neighbor has, where war and physical violence has become socially unacceptable behavior and people's differences are at least tolerated, if not enjoyed.

The appearance of free energy technology in the public domain is the dawning of a truly civilized age. It is an epochal event in human history. Nobody can take credit for it. Nobody can get rich on it. Nobody can rule the world with it. It is simply, a gift from God. It forces us all to take responsibility for our own actions and for our own self-disciplined self-restraint when needed.

**The world as it is currently ordered, cannot have free energy technology without being totally transformed by it into something else.**

This civilization has reached the pinnacle of its development, because it has birthed the seeds of its own trans-formation. Unspiritualized humans cannot be trusted with free energy. They will only do what they have always done, which is take merciless advantage of each other, or kill each other and themselves in the process.

If you go back and read Ayn Rand's Atlas Shrugged or the Club of Rome Report, it becomes obvious that the wealthiest families have understood this for decades. Their plan is to live in the world of free energy, but per-manently freeze the rest of us out. But this is not new. Royalty has always considered the general population (us) to be their subjects. What is new, is that you and I can communicate with each other now better than at anytime in the past. The Internet offers us, the fourth force, an opportunity to overcome the combined efforts of the other forces preventing free energy technology from spreading.

## The Opportunity

What is starting to happen is that inventors are pub-lishing their work, instead of patenting it and keeping it secret. More and more, people are giving away infor-mation on these technologies in books, videos and websites. While there is still a great deal of useless in-formation about free energy on the Internet, the avail-ability of good information is rising rapidly. Check out the list of websites and other resources at the end of this article.

It is imperative that you begin to gather all of the infor-mation you can on real free energy systems. The rea-son for this is simple. The first two forces will never allow an inventor or a company to build and sell a free energy machine to you! The only way you will ever get one is if you, or a friend, build it yourself. This is exactly what thousands of people are already quietly starting to do. You may feel wholly inadequate to the task, but start gathering information now. You may be just a link in the chain of events for the benefit of others. Focus on what you can do now, not on how much there still is to be done. Small, private research groups are working out the details as you read this very message today. Many are committed to publishing their results on the Internet.

**All of us constitute the fourth force.** If we stand up and refuse to remain ignorant and actionless, we can change the course of history. It is the aggregate of our combined action that can make a difference. Only the mass action that represents our consensus can create the world we want. The other three forces will not help us put a fuelless power plant in our basements. They will not help us be free from their manipulations. Nevertheless, free energy technology is here. It is real, and it will change everything about the way we live, work and relate to each other. In the last analysis, free energy technology obsoletes greed and the fear for survival. **But like all exercises of spiritual faith, we must first manifest the generosity and trust in our own lives.**

The source of free energy is inside of us. It is that excitement of expressing ourselves freely. It is our spiritually guided intuition expressing itself without distraction, intimidation or manipulation. It is our openheartedness. Ideally, the free energy technologies underpin a just society where everyone has enough food, clothing, shelter, self-worth, and the leisure time to contemplate the higher spiritual meanings of life. **We owe it to each other to face down our fears and take action to create this future for our children's children?**

Free energy technology is here. It has been here for decades! Communications technology and the Internet have torn the veil of secrecy off of this remarkable fact. People all over the world are starting to build free energy devices for their own use. The bankers and the governments do not want this to happen, but cannot stop it. There will be essentially no major media coverage of what is going on. Tremendous economic instabilities and wars will be used in the near future to distract people from joining the free energy movement.

Western society is in many ways spiraling down toward self-destruction due to the accumulated effects of long-term corruption and greed. The general availability of free energy technology cannot stop this trend. It can only reinforce it. If, however, you have a free energy device, you may be better positioned to support the political/social/economic transition that is underway. The question is, who will ultimately control the emerging world government — the first force or the fourth force?

The last great war is almost upon us. The seeds are planted. After this will come the beginning of a real civilization. Some of us who refuse to fight will survive to see the dawn of the world of free energy. We challenge you to be among the ones who try.

# LIST OF RESOURCES:

## Books:

*Living Energies* by Callum Coats

*The Free Energy Secrets of Cold Electricity* by Peter Lindemann, D.Sc.

*Applied Modern 20th Century Aether Science* by Dr. Robert Adams

*Physics Without Einstein* by Dr. Harold Aspden

*Secrets of Cold War Technology* by Gerry Vassilatos

*The Coming Energy Revolution* by Jeane Manning

## Websites:

http://www.free-energy.cc/ developed by Clear Tech, Inc. and Dr. Peter Lindemann

http://jnaudin.free.fr/ developed by JLN Labs in France

http://www.keelynet.com/ developed by Jerry Decker in the USA

http://www.xogen.ca/ site for super electrolysis technology

http://www.fortunecity.com/greenfield/bp/16/content1.htm site by Geoff Egel, Australia

## Links to other excellent resources:

http://www.WantToKnow.info/resources#newenergy

**Free Energy News Articles — Breakthroughs in major media that should have been headline news:**

Below are verbatim quotes taken from articles at links provided dealing with free energy

**Kids Build Soybean-Fueled Car**, February 17, 2006, CBS News <u>http://www.cbsnews.com/stories/2006/02/17/eveningnews/main1329941.shtml</u>

The star at last week's Philadelphia Auto Show wasn't a sports car or an economy car. It was a sports-economy car-one that combines performance and practicality under one hood. The car that buyers have been waiting decades [for] comes from an unexpected source and runs on soybean bio-diesel fuel to boot. A car that can go from zero to 60 in four seconds and get more than 50 miles to the gallon would be enough to pique any driver's interest. So who do we have to thank for it. Ford? GM? Toyota? No-just...five kids from the auto shop program at West Philadelphia High School.

**Iceland the First Country to Try Abandoning Gasoline**, January 18, 2006, ABC News <u>http://abcnews.go.com/WNT/story?id=1518556</u>

Iceland has already started...turning water into fuel - hydrogen fuel. Here's how it works: Electrodes split the water into hydrogen and oxygen molecules. Hydrogen electrons pass through a conductor that creates the current to power an electric engine. Hydrogen fuel now costs two to three times as much as gasoline, but gets up to three times the mileage of gas, making the overall cost about the same. As an added benefit, there are no carbon emissions - only water vapor.

**Fuel's paradise? Power source that turns physics on its head**, November 4, 2005, The Guardian (one of the UK's leading newspapers) http://www.guardian.co.uk/science/story/0,3605,1627424,00.html

It seems too good to be true: a new source of near-limitless power that costs virtually nothing, uses tiny amounts of water as its fuel and produces next to no waste. Randell Mills, a Harvard University medic who also studied electrical engineering at Massachusetts Institute of Technology, claims to have built a prototype power source that generates up to 1,000 times more heat than conventional fuel. "We've got 50 independent validation reports, we've got 65 peer-reviewed journal articles," he said. "We ran into this theoretical resistance and there are some vested interests here.

**Magnetic energy? Perhaps**, September 7, 2005, San Francisco Chronicle http://www.sfgate.com/cgi-bin/article.cgi?f=/c/a/2005/09/07/BUG9NEJD3L1.DTL

"All we know is that we're seeing more energy output than input. Does Goldes realize what's he's saying — that he's perhaps discovered a clean, inexhaustible energy source? "That's exactly what it appears to be," he answered. A handful of other companies worldwide are believed also to be pursuing zero-point energy via magnetic systems. One of them...is run by a former scientist at NASA's Jet Propulsion Laboratory in Pasadena. According to Aviation Week & Space Technology magazine, the Pentagon and at least two large aerospace companies are actively researching zero-point energy as a means of propulsion.

**Solar Challenge Finishes in Calgary**, July 28, 2005, Open Source Energy Network/Detroit News http://pesn.com/2005/07/28/ 9600141 Solar Challenge results http://www.detnews.com/2005/schools/0507/28/01-262474.htm Detroit News http://wwmt.com/ engine.pl?station=wwmt&ampid=18269&amptemplate=breakout_local.html - CBS affiliate

The ten-day solar car race from Austin to Calgary came to a successful finish yesterday. U of Michigan takes prize, finishing the 2500-mile course in 54 hours. They also set a record by averaging 46.2 mph in this, the world's longest solar car race.


**Eco-car more efficient than light bulb**, July 5, 2005, CNN http://www.cnn.com/2005/TECH/07/04/eco.car

The hydrogen-powered Ech2o needs just 25 Watts — the equivalent of less than two gallons of petrol — to complete the 25,000-mile global trip, while emitting nothing more hazardous than water. But with a top speed of 30mph, the journey would take more than a month to complete. Ech2o, built by British gas firm BOC, will bid to smash the world fuel efficiency record of over 10,000 miles per gallon at the Shell Eco Marathon. The record is currently....5,385 km/per liter [over 12,000 mpg!].

**Advanced vehicles demonstrate zero oil-con-sumption**, reduced emissions May 18, 2005, Boston G l o b e http://www.evworld.com/ view.cfm?section=communique&newsid=8474 (article removed from Globe website)

Top prize for the Monte-Carlo Rally went to a modified Honda Insight [which] broke the 100-mile-per-gallon barrier over a 150-mile range. The car actually got 107 miles-per gallon. St. Mark's High School in Southboro, and North Haven Community School, North Haven, ME, demonstrated true zero-oil consumption and true zero climate-change emissions with their modified electric Ford pick-up and Volkswagen bus.

**Fans of GM Electric Car Fight the Crusher**, Washington Post, March 10, 2005 http://www.washingtonpost.com/ac2/wp-dyn/ A21991-2005Mar9

GM agrees that the car in question, called the EV1, was a rousing feat of engineering that could go from zero to 60 miles per hour in under eight seconds with no harmful emissions. The market just wasn't big enough, the company says, for a car that traveled 140 miles or less on a charge before you had to plug it in like a toaster. Ted Flittner, a...Costa Mesa industrial engineer...said, "they have such a brilliant solution they've developed. They've put it on the market and proved it works. People still want it and they're taking it away and destroying it."

**100 MPG Car Heralded by London Times in 2002 - Where is it now?** December 2 , 2004, WantToKnow.info/ London Times http://www.WantToKnow.info/carmileage - WantToKnow.info (includes text of London Times article) http://www.timesonline.co.uk/article/0,,588-451038,00.html - London Times

The Toyota Eco Spirit was the talk of the fuel economy car industry in 2002. At over 100 MPG and with the lowest exhaust emissions and a very reasonable sticker price, the Eco Spirit's debut was widely anticipated. (see London Times article). What happened to it? 1908 Ford Model T: 25 MPG, 2004 EPA Average All Cars: 21 MPG Detroit News/WantToKnow.info, June 4, 2004 http://www.WantToKnow.info/ 050711carmileageaveragempg

Ford's Model T, which went 25 miles on a gallon of gasoline, was more fuel efficient than the current Ford Explorer sport-utility vehicle — which manages just 16 miles per gallon.

Note: This last article is an excellent summary of eye-opening contradictions which have received very little media coverage, including links to major media articles to back up the facts presented.

**Patents: This list is a sample of inventions that produce free energy.**

To search for patents by number on the US Patent Office website:
**http://patft.uspto.gov/netahtml/PTO/srchnum.htm**

| | |
|---|---|
| Tesla | USP # 685,957 |
| Freedman | USP # 2,796,345 |
| Richardson | USP # 4,077,001 |
| Frenette | USP # 4,143,639 |
| Perkins | USP # 4,424,797 |

Hope Deferred

# Appendix

Hope Deferred

## Eco friendly flying car

Sounds pretty un eco friendly, but it is supposed to be green even though ***it has eight wankel style engines***, reaches a top speed of 50 mph and has a 900 mile range limit. There are plans for a more comfortable enclosed flying car/plane with just four engines.

But this flying saucer type machine is the one that many people are interested in, the company plans to build around 250 of these vehicles a year, which will even need the driver or pilot to have a pilots licence because the machine will only fly at about nine feet off the ground. Some people do not like the idea of one these in the hands of a drunk driver, because these flying cars could cause an untold amount of damage.

# Nikola Tesla Predicted Smartphones 100 Years Ago.

While some may think that the plethora of smartphones and other handheld devices may seem futuristic and ever so modern; the fact is, visionaries have been imagining and predicting similar technology for more than 100 years.

Nikola Tesla, one of the fathers of the modern alternating current (AC) electrical system we use everyday, stated in an interview with New York Times back in 1909 that in the future it would be possible to send wireless messages back and forth across the globe. He envisioned doing so with an easy-to-use handheld device that would readily be available to the masses.

---

### WIRELESS OF THE FUTURE

(Nikola Tesla in the New York Times)

"It will soon be possible, for instance, for a business man in New York to dictate instructions and have them appear instantly in type in London or elsewhere. He will be able to call up from his desk and talk with any telephone subscriber in the world. It will only be necessary to carry an inexpensive instrument not bigger than a watch, which will enable its bearer to hear anywhere on sea or land for distances of thousands of miles. One may listen or transmit speech or song to the uttermost parts of the world. In the same way any kind of picture, drawing, or print can be transferred from one place to another. It will be possible to operate millions of such instruments from a single station."

---

Sound a bit familiar?

He proposed that this wireless messaging wave to bring along with it, an entirely new era of technology. Predicting the coming of "wireless power," which is in its infancy today but is already in use in products like Powermat's charge pads for handheld devices including iPhone, Android and Blackberry smartphones.

*Nikola Tesla* demonstrated the wonders of alternating current electricity in Chicago in 1893, which became the standard power in the 20th Century. Tesla designed the first hydroelectric powerplant in Niagra Falls in 1895. The Tesla coil, which he invented in 1891, is widely used today in radio and television sets and other electronic equipment. Among his discoveries are the fluorescent light; laser beam, wireless communications, **wireless transmission of electrical energy,** remote control & robotics. Tesla is the father of the radio and the modern electrical transmissions systems. Registering over 700 patents world wide, his vision included exploration of solar energy and the power of the sea. He foresaw interplanetary communications and satellites.

Hope Deferred

# Other Publications

NESARA: National *Economic Security
and Reformation Act*
http://tinyurl.com/c8u42q6

History of Banking: *An Asian Perspective*
http://tinyurl.com/boeehjl

The People's Voice: *Former Arizona
Sheriff Richard Mack*
http://tinyurl.com/d62fyg3

Asset Protection: *Pure Trust Organizations*
http://tinyurl.com/btrjfqp

The Matrix As It Is: *A Different Point Of View*
http://tinyurl.com/ckrbkge

From Debt To Prosperity: *'Social Credit' Defined*
http://tinyurl.com/d2tjmw3

Give Yourself Credit: *Money Doesn't Grow On Trees*
http://tinyurl.com/d7tphuv

My Home Is My Castle: *Beware Of The Dog*
http://tinyurl.com/bmzxc2n

Commercial Redemption: *The Hidden Truth*
http://tinyurl.com/d9etg7w

Hardcore Redemption-In-Law: *Commercial Freedom And
Release*
http://tinyurl.com/cl65vrz

Oil Beneath Our Feet: *America's Energy Non-Crisis*
http://tinyurl.com/btlzqxf

Untold History Of America: *Let The Truth Be Told*
**http://tinyurl.com/bu9kjjc**

Debtocracy: *& Odious Debt Explained*
**http://tinyurl.com/cooqzuz**

New Beginning Study Course: *Connect The Dots And See*
**http://tinyurl.com/cxpk42p**

Monitions of a Mountain Man: *Manna, Money, & Me*
**http://tinyurl.com/cusgcqs**

Maine Street Miracle: *Saving Yourself And America*
**http://tinyurl.com/d4yktlw**

Reclaim Your Sovereignty: *Take Back Your Christian Name*
**http://tinyurl.com/cf5taxh**

Gun Carry In The USA: Your Right To Self-defence
**http://tinyurl.com/cdn3y3y**

Climategate Debunked: *Big Brother, Main Stream Media*
**http://tinyurl.com/d6gy2xz**

Epistle to the Americans I: *What you don't
know about The Income Tax*
**http://tinyurl.com/d99ujzm**

Epistle to the Americans II: *What you don't
know about American History*
**http://tinyurl.com/cnyghyz**

Epistle to the Americans III: *What you don't
know about Money*
**http://tinyurl.com/cp8nrh8**

Hope Deferred

www.ingramcontent.com/pod-product-compliance
Lightning Source LLC
Chambersburg PA
CBHW051333170526
45166CB00002B/802